Work Organization and Methods Engineering for Productivity

Work Organization and Methods Engineering for Productivity

D.R. Kiran

BSP BOOKS

ELSEVIER

Butterworth-Heinemann
An imprint of Elsevier

Butterworth-Heinemann is an imprint of Elsevier
The Boulevard, Langford Lane, Kidlington, Oxford OX5 1GB, United Kingdom
50 Hampshire Street, 5th Floor, Cambridge, MA 02139, United States

British Library Cataloguing-in-Publication Data
A catalogue record for this book is available from the British Library

Library of Congress Cataloging-in-Publication Data
A catalog record for this book is available from the Library of Congress

ISBN: 978-0-12-819956-5

For Information on all Butterworth-Heinemann publications
visit our website at https://www.elsevier.com/books-and-journals

Publisher: Mathew Deans
Acquisitions Editor: Brian Guerin
Editorial Project Manager: Mariana L. Kulh
Production Project Manager: Anitha Sivaraj
Cover Designer: Victoria Pearson

Typeset by MPS Limited, Chennai, India

Contents

About the author

Prof. D.R. Kiran, B.Sc., B.E., M.Sc. (Eng.), (Ph.D.), FIE(I), FIIIE, FIIProdE, FIIPlantE, FITTE, FISNT, has a rich practical experience of forty years both in industry and academy. Starting his career in 1968 with Larsen & Toubro, he held top positions like World Bank adviser/instructor for Transport Managers in Tanzania, planning manager of Rallifan (CF division) and the principal of a Chennai based Engineering College. He has been in the doctors' panel for a Ph.D. student in 2008.

In recognition of his services in the field of engineering education, he was presented with the coveted Bharat Excellence Award and Gold Medal for Excellence in Education in New Delhi in 2006. He is listed as an International Expert in Industrial Engineering and Management in the International Directory of Experts and Expertise. He is nominated for the post of Honorary Deputy Director General in India for International Biographical Center.

Earlier during the eighties, he was introduced to Dr. Julius Nyerere, then the president of Tanzania as a pioneer of Work Study in that country. He was one among few non-political foreigners to be interviewed by the government newspaper of Tanzania.

He started his academic career in 1979, and taught subjects like Work Study, Production Planning & Control, Total Quality Management, Professional Ethics and Maintenance Engineering Management at B.E. level.

He was responsible for the establishments of many student chapters for IIPE and NIQR. His active participation in these activities and his experience as the planning manager of Rallifan and National Bicycle Company where he was the overall in charge of the PP&C department, provided him the inspiration for him in planning for a comprehensive book on Principles and Practice of Work Study, providing several case studies on key topics out of his rich experience. This is expected to be of immense help to the students as well as the practicing engineers.

His career as an industrial engineer and as planning manager, heading the function of IE as well as PP&C at Rallifan and National Bicycle Co. gave him an opportunity to introduce several systems and procedures in production management. Having been qualified in Industrial Engineering, he conducted several method study projects both as an internal and external consultant. The absence of a comprehensive Indian book with case studies motivated him to author this book on "*Work Organization and Methods Engineering for Enhanced Productivity*", which is his fifth text book for engineering students.

Earlier he has authored four engineering text books referred both at BE and PG level,

- Professional Ethics and Human Values, published by McGraw Hill Education, India,
- Maintenance Engineering and Management - Precepts and Practices,
- Total Quality Management, Key Concepts and Case Studies and
- Production Planning and Control, A Comprehensive Approach.

While the TQM and PP&C books have been published in USA by Elsevier. The Maintenance Management book has been published in USA by Taylor & Francis. All the three were launched successfully in Boston.

He had also published over 23 papers in Indian and Foreign Journals as well as in conference Proceedings. He was the organizing secretary for the successful 29th Production Convention of the Institution of Engineers held in August 2014. He is widely traveled having visited over 30 countries and is a philanthropist. Having taught engineering students for over 3 decades, he demonstrated his love for the student community by donating for the annual best student awards at IE(I), ISNT, at Venkateswara Vidyalayam.

Preface

Productivity, rather increase in productivity is the topic of the day and work study is tailored towards that goal. The term work study is not new to industry sector. It is a technique combining two basic tasks. The first one is that of organizing the work place layout and measuring the work content, both from the ergonomic and productivity point of view or in other words, the work organization. The other task is that of developing optimal methods of operation from the productivity point of view, or in other words, the methods engineering. These two wings of work study are integrated and refined into a widely accepted technique applicable to the improvement and upgrading of work systems. This book is hence aptly titled '*Work Organization and Methods Engineering for Enhanced Productivity*', though in the book, it is generally referred to by its more popular synonym, the Work Study.

Although the second decade of the 21st century is experiencing a lot of disruptive advancement in industrial operations by Artificial Intelligence and Internet of Things, the basic understanding of analyzing the human motions and organizing the work place with reference to the productivity improvement is still pertinent. An attempt is hence made to present the subject matter in a reasonably simplified manner, enriching the text with illustrations and case studies. The topics are discussed with more general knowledge than technical knowledge in mind so that even those who do not have much exposure to Work Organization function through work study practice would be able to comprehend them.

While a majority of case studies discussed in this book refer to the manufacturing industry, other related fields like the textile mills and sea port operations are illustrated. As the head of both the Industrial Engineering as well as the Production Planning functions at Ralli group of industries, the author's experience has fully been utilized in bringing out this book on the basics of work study with several formats and charts that are useful in recording the data.

A feature of this book is the detailed discussions on work sampling, ergonomics, role of consultants, etc. In view of Japan becoming a strong industrialized nation during the later half of 20th century, it has become a fashion among Indian, rather the global managers, to use several Japanese management terms in their day to day work. One chapter of this book is hence devoted to explain the most commonly used Japanese management terms.

Another feature of the book is the introduction of a new recording technique called *time-lapse camera video recording*, as explained in Chapter 7. Perhaps this is the first text book on work study to represent this recording technique.

It is ensured that every chapter precedes with chapter synopsis and key-words used, so as to provide the reader a first glimpse of the chapter contents and highlights. In addition to the chapter oriented references given in each chapter, a bibliography is included at the end of the book. Every chapter ends with criteria questions giving paragraph clues, from which the answer can be located. Syllabi from several Universities and Institutions of India as well as International Universities are collected to draw the outline for this book.

Prof. D.R. Kiran

Acknowledgments

The author wishes to acknowledge his indebtedness to all the persons who were associated with him during his 45 year-long industrial and academic careers, and who helped him in bringing the book '*Work Organization and Methods Engineering for Enhanced Productivity*' to this shape.

Useful computer software

Work measurement (Chapter 14)

1. WorkStudy+
2. Umt Plus
3. UMT Manager
4. UmtPlus (for PDA)
5. StatUmt
6. StandardsPro

Micro motion study (Chapter 15)

1. MOST Computer Systems
2. Basic MOST or Mini MOST
3. Maxi MOST
4. Clerical MOST
5. Admin MOST

Ergonomics and work study (Chapter 16)

1. Ergopoint
2. Medgate's Ergonomics Software
3. JET Ergonomic Assessment Software
4. Intelex Ergonomics Analysis Software
5. 3DSSPP Software
6. Ergoweb Enterprise

Work sampling (Chapter 17)

1. PATHSoft
2. Laubrass' UMT Plus Software
3. The Work Sampling System

Abbreviations

3 Mu	Muda, Muri and Mura
4 M	Man, Machine, Material and Method
5S	SEIRI, SEITON, SEISO, SEIKETSU and SHITSUKE
5 Zu's	Uketorozu, Tsurazu, Baratsukasazu, Kuriakalsazu and Nagasazu
AMA	American Management Association
BOM	Bill of Materials
BT	Basic Elemental Time, also Budgeted Operational Time
CAD	Computer Aided Design
CAM	Computer Aided Manufacturing
CIEHS	Chartered Institute of Ergonomics and Human Factors (UK)
CPU	Central Processing Unit
CREW	Cost Reduction through Elimination of Waste
CRS	Cold Rolled Steel
CSVA	Canadian Society of Value Analysis
DARSIRI	Data (collect), Analyze, Record ideas, Speculate, Innovate, Review and Implement
DFSS	Design for Six Sigma
DMADV	Design, Analyze, Measure, Define and Verify
FA	Fatigue Allowance
FA	Functional Analysis
FAST	Functional Analysis System Technique
HFES	Human Factors and Ergonomics Society (USA)
HMI	Human-Machine Interaction
ICH	International Council for Harmonization
ID	Inside Diameter
IEA	International Ergonomics Association
ILO	International Labour Organization
IMD	International MTM Directorate
INVAVE	Indian Journal of Value Analysis/Value Engineering.
INVEST	Indian Value Engineering Society
IVM	UK Institute of Value Management
JIT	Just in Time
JITB	Joint Industrial Training Board
JUSE	Japanese Union of Scientists and Engineers
MA	Miscellaneous Allowance
MIS	Management Information System
MMMM (4 M)	MTM – 4th Generation

MOST	Maynard Operation Sequence Technique
MSD	Musculo-Skeletal Disorders
MTM	Methods-Time Measurement
MTM −1	Methods-Time Measurement − 1
MTM- 2	Methods-Time Measurement − 2
MTM- 3	Methods-Time Measurement − 3
MTMASR	MTM Association for Standards and Research
NPTEL	National Programming on Technology Enhanced Learning
NSIC	National Small Industries Corporation
O&M	Organization & Methods
OD	Outside Diameter
OEEC	Organization for European Economic Cooperation
OPC	Outline Process Chart (refer Para 7.5.3)
OT	Observed Operational Time
PMTS	Predetermined Motion Time Standards
PP&C	Production Planning and Control
RF	Rating Factor
SAVE	Society of American Value Engineers
SIDCO	Small Industries Development Co-operation
SIMO	chart Simultaneous Motion Chart
SJVE	Society of Japanese Value Engineering
SMED	Single Minute Exchange of Die
SOP	Standard Operating Procedure
SPC	Statistical Process Control
SREDDIM	Select, Record, Evaluate, Develop, Define, Install and Maintain
ST	Standard Operation Time
TISCo	Tata Iron and Steel Company
TLCV	Time-Lapse Camera Video
TMU	Time Measurement Unit
TPM	Total Productivity Measure
TPM	Total Productive Maintenance
VA/VE	Value Analysis and Value Engineering
VM	Value Management
VMd	Value Methodology
WRMSDs	Work-Related Musculoskeletal Disorders

Chapter 1

Introduction

1.1 Work organization and methods engineering

Work Organization and Methods Engineering is a subspecialty of industrial engineering. While Work Organization is concerned with human integration in industrial production processes by skills distribution and coordination of work tasks, Methods Engineering is the analysis and development of the method being employed in performing these tasks to achieve the objective of lowering the production costs and increasing reliability and productivity. Work study is that branch of industrial engineering, that aims at achieving both these objectives. For this reason, we use the term work study in this chapter as well as the other chapters of this book, to represent both Work Organization and Methods Engineering.

1.2 What is work study

Before we attempt to answer this question, let us review some of the explanatory definitions available on work study, which may possibly outline what it incorporates. It basically is a system of assessing methods of working so as to achieve the maximum output and efficiency. Work study helps to increase productivity of men, machines and materials.

IS 6363:1972 (Glossary of terms used in work study) of the Bureau of Indian Standards, defines work study as

Work study is a modern discipline, which analyzes and evaluates all aspects of the work system in order to enhance the effectiveness and functional efficiency.

BS 3138:1979 (Glossary of terms used in work study) of the British Standards Institution gives a more typical and comprehensive definition, which is more an explanatory definition as

> *Work-study is a generic term for those techniques, particularly of method study and work measurement, which are used in the examination of human effort in all its contexts and which leads systematically to the investigation of all factors that affect the efficiency and economy of the situation being reviewed in order to affect the improvement.*

Work Organization and Methods Engineering for Productivity.
DOI: https://doi.org/10.1016/B978-0-12-819956-5.00001-7

R.H. Hammond in his chapter History and Development of Industrial Engineering in the Production Hand Book edited by Gordon Carson, refers to the Report on the Training for Work Study Practices of the Joint Industrial Training Board, which states as follows

> *Work study attains its benefits through, firstly by investigation of the current situation, examining especially any apparent weaknesses, for example, the performance of an operating team or a machine group. This diagnosis is followed by the determination and the introduction of appropriate improvements in operating methods. Then investigating and review will cover operating methods, selection and usage of equipment, plant layout supply and usage of materials, availability of ancillary services like material handling, work organization, effectiveness of total operating procedures, progress control and the potential effect of the investigations on overall costs and efficiency.*

The Free dictionary defines work study as the analysis of industrial or work procedures to determine the most efficient methods of operation.

We may cite a hoard of such explanatory definitions, which have one thing common with them - no single sentence definition can fairly and adequately explain the subject as much as the above definitions, or rather explanatory definitions do. They reveal how boundless work study can be in the day to day works in all walks of life, whatever it is a manufacturing industry or an automobile workshop, or a chemical industry or even an educational institution.

1.3 Work study is a tool for increasing productivity

From the above definitions, we can see that work study has a direct relationship to productivity and is the most frequently used technique for increasing the amount produced (output) from a given resource (input), with little or no increase in the capital investment. Productivity, of course can be increased in the long run by innovations and development of new processes, modernizing of plant and equipment or by acquisition of advanced technology. But these need heavy capital outlay and cause a drain of our meager foreign exchange and other resources. No one would like to increase the capital outlay just like that. This aspect is discussed more in detail in Chapter 11 on kaizen.

We shall hence look at the productivity problem from a different angle. Let us determine the causes of low productivity by systematic analysis of the existing processes, designs and work methods, thereby devising means and methods to increase output by eliminating or modifying wasteful elements of operations, designs etc., with no or minimal increase in the capital. But after all, history has time and again shown that, what is supposedly impossible in the yesteryears has become a possibility today, whether it is a steam engine or man landing on the moon or the desk-top and lap-top computers, or more so, in smart machines or IoT.

It may be true that some improvements like the discovery of steam power by James Watt had come by pure chance. Nevertheless, many of the successes achieved in the recent past have been results of systematic applications of the work study principles. In these days of scientific and technical advancement, it would indeed be an unforgivable crime to rely on pure chance and here lies the need to have some reliable predetermined knowledge of the implications and consequences of a change.

After all, investigations and improvements of operations or increase in productivity are not new in the history of industrial development. Many geniuses starting from Frederick Taylor have made notable advances towards an increase in productivity. Unfortunately, we are yet to come across a factory that would produce such geniuses in a mass production line to cater for the highly complex industrial situations of today. Thus, it is of prime importance that the process of improvements be made so systematic, foolproof and at the same time, simple that any normal manager, by carrying out this systematic procedure can achieve results as good as, or better than what a less systematic genius was able to achieve in the past.

1.4 Work study is a tool to achieve production objective

The basic objective of production management is to produce the right quantity of goods in right quality at a predetermined time at a predetermined cost. Work study is the tool with which the management strives to achieve this objective by providing standard methods of operation to the manufacturing activities.

The prime value of work-study lies in the fact that if a systematic procedure is applied both in the investigation into the problem and the development of solutions, one is sure that no criteria or limitation or alternative is left unconsidered and can confidently say that he had left no stone unturned. This is a prerequisite for effective results and this aspect undoubtedly distinguishes a systematic application of work study from chance improvements.

1.5 Work study is a penetrating tool of investigation

As we have defined earlier, work study is concerned with the analyzes for the optimization of complex processes, systems or organizations, leading to the elimination of waste of time, money, materials, man-hours, machine time, energy and other resources that do not generate value. This is why it is also referred to as CREW (Cost reduction by elimination of waste). It hence involves investigations by direct observations of all the factors affecting the efficiency of an operation and is bound to show up any shortcomings in all the activities of an organization. A common example is the low productivity due to the forced idleness caused factors, frequent machine breakdown or non-availability of the raw material at the right time in the right place in the right quantities. In ILO's expression, work study acts like a surgeon's knife laying bare the activities and functioning of all the different departments of an origination, good or bad, for all those to see. These point inaction by respective officials in activities like maintenance and material control. That means it shows up people, and obviously no one likes to be shown up.

In other words, what we may start as a simple methods improvement may reflect on the effectiveness of one or more senior executives, and may well cause considerable resentment and in some cases, active opposition by the aggrieved, which might boomerang in situations where the work study engineer lacks adequate managerial support. Basically, such an opposition or resistance is caused due to the fact that many executives view such an exposition as an isolated case of loss of their prestige rather than by viewing it from the general overall benefit point of view. For this reason, work study must be handled like a surgeon's knife, with all the skill and care that are needed.

Therefore, it is vital that all concerned and involved in the implementation of the change must be taken into confidence while conducting the study. Secrecy or the appearance of secrecy must be avoided at all costs. The ideal situation is one where work study forms an integral part of all activities and not a separate entity. Chapter 20 may be referred to, wherein various aspects resistances change which is inherent to human nature and discusses how the work study engineer should strive to overcome the same.

1.6 Work study and productivity

As seen in the previous paragraphs, productivity implies the development of an attitude of mind and a constant urge to find better, cheaper, easier, quicker and safer means of doing a job, manufacturing a product or providing a service.

The whole principle of various industrial engineering techniques including that of work study is reflected in the above statement.

In other words, work study emphasizes in identifying, analyzing, and critically examining various factors that add to the ineffective time involved in the production of a product. By systematic procedure, detailed in subsequent chapters, alternative methods or processes are developed whereby productivity is effectively improved. This aspect is dealt more in detail in Chapter 3.

1.7 Work study is an unquestioned valuable tool for the management

We can summarize the reasons for considering work study as a valuable tool of management as

1. It is a direct means of raising productivity and production efficiency of an organization or an operating unit, involving little or no capital expenditure on plant and equipment.
2. It is based on systematic, consistent and simple principles and procedures where analysis is based on fact and data and not on individual opinions.
3. It ensures that no factor or data affecting the efficiency are overlooked.
4. The various recording techniques adapted help in presenting all the facts as well as the analysis in simple but clear cut and specific charts, symbols and reports, so that not only the bosses, but personnel at all levels who are involved in the implementation of the change, can understand and appreciate the procedure adopted.
5. It is a reliable and accurate means of setting up targets of performance, which help effective planning and control of not only the production but also other activities like maintenance.
6. An effectively conducted work study results in immediate savings to the company and would continue as long as the change continues.
7. Work study has universal applications. It can be used wherever the manual work is involved, whether it is machining operation or transport, workshop, clerical work, hospital, supermarket or even educational institution.
8. As explained in paragraph 1.4, work study is a penetrating tool of investigation, and shows up all the ineffective practices by the staff at different levels, as well as the invisible ineffective elements, similar to the bottom of the iceberg hidden in the ocean, as illustrated in Fig. 4.1.

1.8 Work study is an interdepartmental function

Like any other functional activity that performs an advisory role or as a staff function, work study cannot work in isolation, but forms a central function round which all other departments stand to benefit, as illustrated in Fig. 1.1. It is hence imperative that a regular flow of information to other related departments is indicated in Fig. 1.1 is maintained.

1.9 Inter-departmental information flow

Inter-departmental Communication is one of the organizational functions that help smooth and streamlined flow of information necessary for the uninterrupted production flow; this helps a company to stay efficient and productive. It is the communication between departments that keep the organizations alive and efficient and when it breaks down, crises would develop affecting the smooth production flow. The following illustrations can be cited to highlight the importance of prompt communications between several departments and sections.

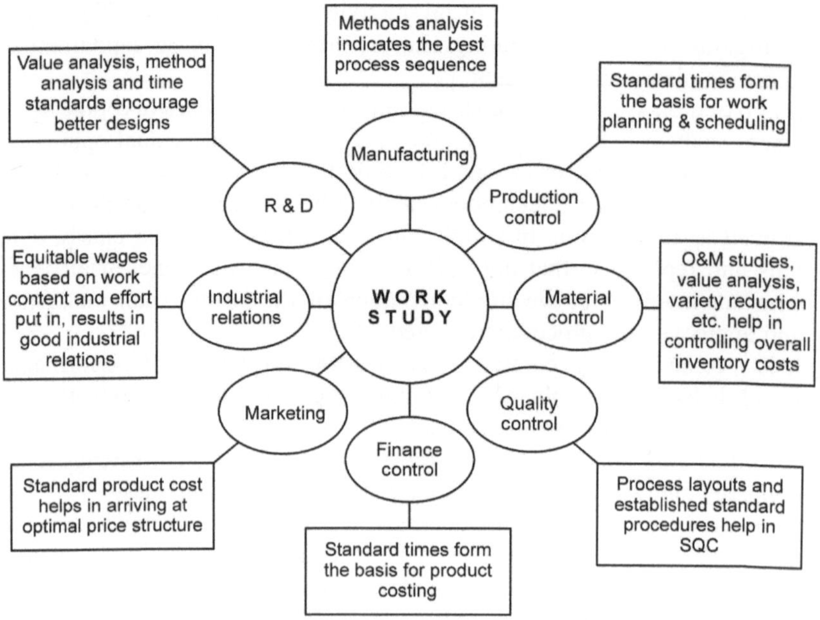

FIGURE 1.1 Relationship between work study and other departments.

1. When inter-departmental communication is poor, customer service too can suffer. For example, if the accounts receivable department is not communicating properly with accounts payable, a client continues to receive a bill for an invoice that was already paid and this will result in losing the customers and the repeat business.
2. If the production department is not appraised about an increase in product demand, then the company suffers a loss of revenue. This would also result in subsequent priority allocation on war footing, resulting in poor efficiency.

We can cite hordes of similar illustrations that we experience in our day to day manufacturing life.

Fig. 1.2 illustrates (See next page) how the policy and the feedback communication flow between several productions and related departments. For further information on Inter-departmental Communication, reference may be made to the book *'Production Planning and Control, A Comprehensive Approach'* by this author.

1.10 Management information systems

Despite conducting an excellent method study project, the work study engineer's efforts would be appreciated only if he makes a good report of the

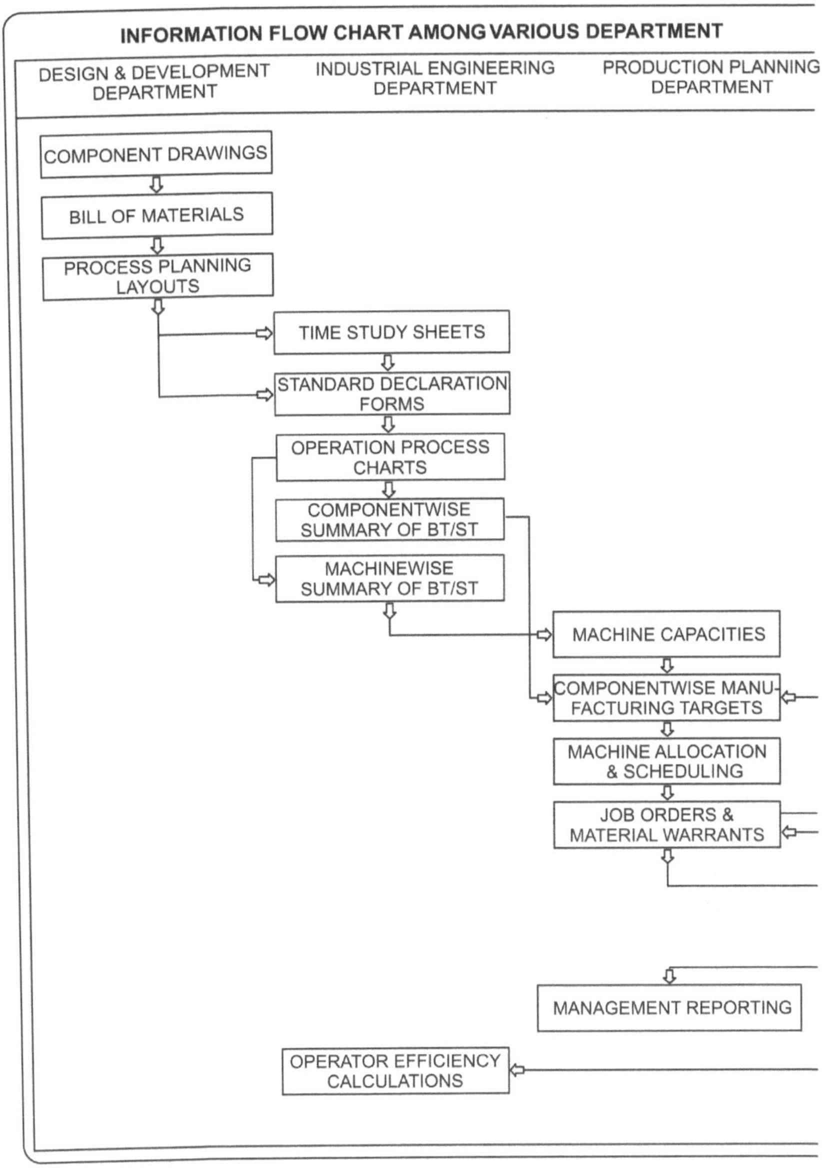

FIGURE 1.2 Information flows between several departments.

work. For this he must understand the principles of management information system, as explained in the next paragraphs.

Feedback system is the communication or an effective flow of information as well as the policy at the right time to the management. In general, for effective communication, the policy information flows from the management

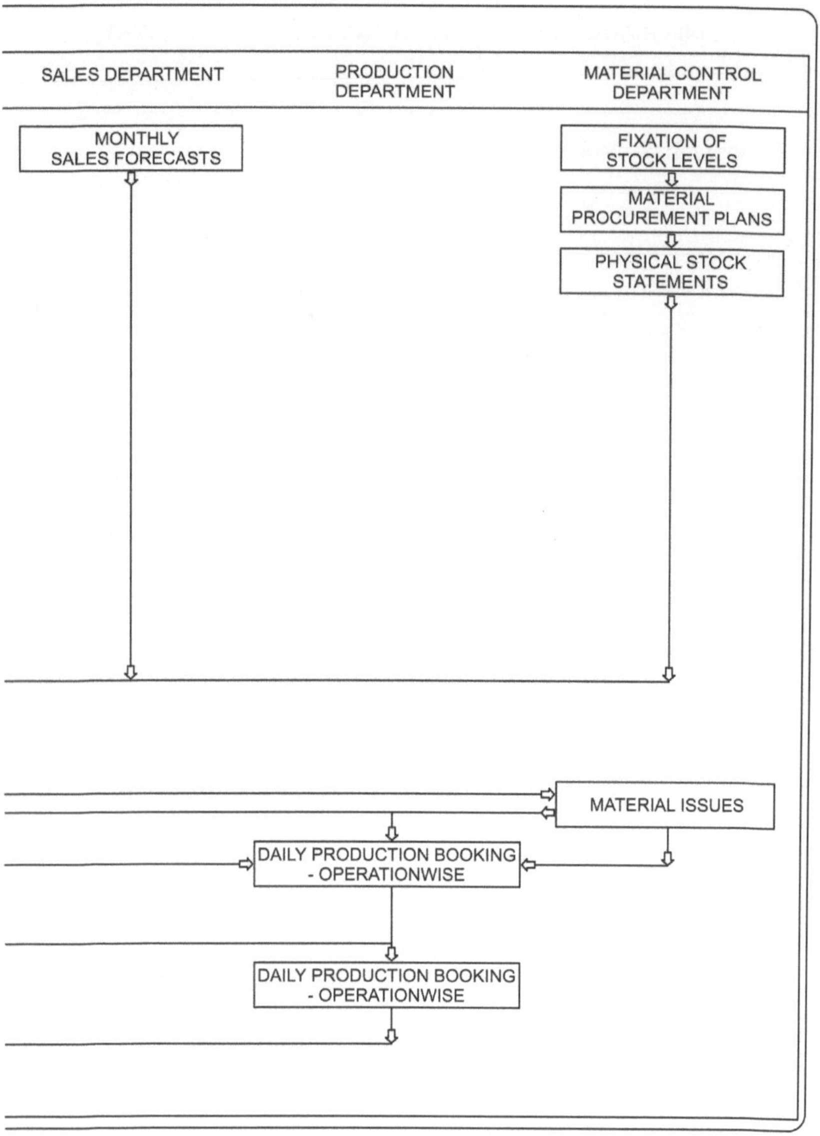

FIGURE 1.2 (Continued).

to the planning and control staff, while the feedback information flows from the control staff to the management. This is represented in Fig. 1.3.

1.10.1 Why MIS is essential for the managers?

When a manager is in the dark about what is happening in the factory, he is in a state of uncertainty and cannot decide his next step. An effective

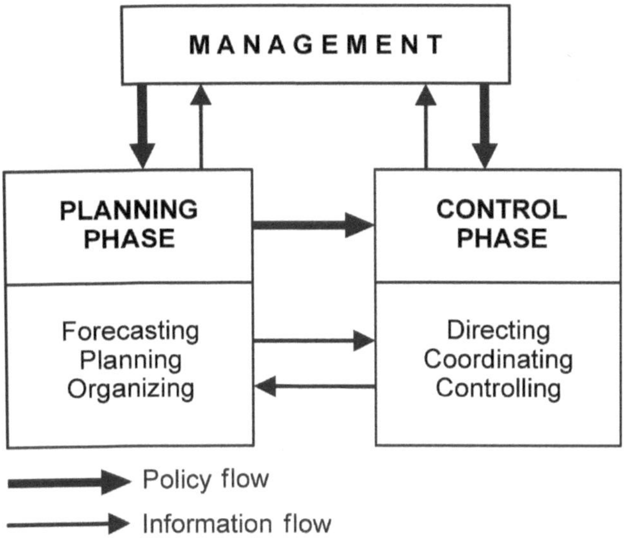

FIGURE 1.3 Management information system.

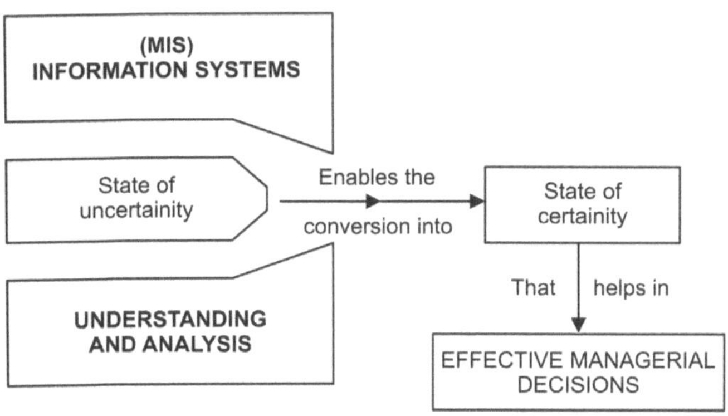

FIGURE 1.4 How effective MIS help in managerial decisions.

information system can help him understand and analyze the situation. In other words, an effective MIS provides facts based on which the manager can take a decision. Fig. 1.4 illustrates how the MIS assists the manager as above. Chapter 21 on the role of consultants, explains MIS further and discusses its significance in making the job of work study engineers and consultants more effective.

1.11 Why we need specialists in work study

We have said that work study is systematic and at the same time is simple to understand. Then why do we need specialists for work study? Can the production personnel not conduct a work study themselves and achieve results?

Well, before we answer this, let us refer to paragraph 1.3 where we said that work study succeeds due to the fact that all the limitations, criteria and alternatives are considered by systematic investigations. These detailed investigations take time and a high amount of concentration of thoughts and specially acquired skills. The production manager or the supervisor might have acquired these skills and is efficient, but he is engrossed in his day-to-day and minute-to-minute technical and other production problems that he is never free for long interruptions from his work of achieving the production targets. It would be much more difficult for him to gather all data patiently and unless all the facts are fully known, it would be difficult to continue investigations.

This means that the function of a work study must always be the responsibility of someone, who is free from production problems like the stress of achieving production targets. He should be able to undertake his job as a staff function and not as a line function.

1.12 Psychological aspects of work study

Management is both an art and science. There are a number of scientific techniques that can be applied to solve management problems with systematic, step-by-step approach from the known to the unknown on the basis of ascertaining facts.

While scientific techniques are applied to materials which are governed by known physical laws, the management techniques have to be applied to people by people. A full understanding of the human behavior, especially of those affected by the decision is essential for the very success of these techniques. We can say that while the systematic analysis forms the science part of management, the human relations form the art part of the management. The chief aim of every management is to lead the enterprise towards specified goals. This can be achieved by organizing and controlling all activities, especially the human activities of the organization.

Despite industrial engineers having widened into most of the management techniques like operations research and analysis, robotics etc., work-study continues to be their chief activity. Industrial productivity aims at method improvement or systems improvement. But the very word 'improvement' is linked to an action of change and the resultant resistance is more psychological that the workers being against the change itself. A sincere attempt to understand the possible motive for such a resistance, as well as the past history responsible for the development of such a resistance, together with the complete analysis of the situation would possibly enable the management to plan a rational course of action for the successful implementation of the change. This aspect is dealt much more in detail in chapter no. 20 on Resistance to change.

1.13 Conclusion

As highlighted in paragraph 1.12, there is no doubt that, even though the industrial engineer's activities have widened into research issues, robotics, Internet of Things etc., their principal activity continues to be work study. Today, even large industries are resorting to stopwatch time study for setting work standards. For this reason, not only industrial engineers, but all those who work in operational controls, should acquire sufficient knowledge about work study, and this book strives towards that direction.

Criteria questions

1. How do you relate the term Work Study with Work Organization and Methods Engineering? (1.1)
2. How do you justify 'Work study is a tool for increasing productivity'? (1.3)
3. How can work study be a penetrating tool of investigation? (1.5)
4. Represent in a sketch, how Work study is related to other functional activities. (1.8)
5. Why we need specialists in work study, free from production responsibilities? (1.10)
6. Distinguish between data and information. (1.11)

Further reading

1. International Labor Organization. *Introduction to work study*. 3rd ed. ILO. 1979.
2. Landen K. Management's use of industrial engineering. In: Maynard HB, editor. *Industrial engineering handbook*, 2nd ed. McGraw Hill.
3. Hammond RH. History and development of industrial engineering. In: Carson G, editor. *Production handbook*, Arnold Press.
4. Kiran DR, editor. *Proceedings of the work study seminar*. University of Dar Es Salaam; 1983.
5. Kiran DR, editor. *Proceedings of the management seminar for transport sector*. Dar Es Salaam: National Institute of Transport; 1984.
6. Kiran DR. *Programmed instructions manual for NCES*. Chennai: NPC; 1985.
7. Kiran DR. Participative management, a case study. *J Indian Inst Prod Eng* 1999.
8. BS 3138. *Glossary of terms used in work study*. Bristish Standards Institute; 1992.
9. IS 6363. *Glossary of terms used in work study*. Bureau of Indian Standards; 1997.
10. Papa MJ, Daniels TD, Spiker BK. *Organizational communication*. Sage Publications; 2007.

Chapter 2

History and development of work study

2.1 Early man is the unacknowledged pioneer in work study

The concept or the urge to do things in an easier and better ways has always been present in man ever since he started living as a social animal, ever since he learnt to eat cooked meat or discovered wheeled carts. Nevertheless, these attempts are very much unsystematic and chance oriented, and hence could not be attributed as steps in the development of new methods in the form of work study.

In fact, the history of work study is closely related to the history and development of scientific management, whose origin can be traced back to the industrial revolution that took place in the 18th and 19th centuries.

2.2 The industrial revolution

The start of the industrial revolution is generally attributed to the period around 1750, corresponding to the reign of Elizabeth in England, when the development of domestic woolen units has accelerated the economic changes. Nevertheless, the more marked period of industrial revolution can really be attributed to the period between 1815 and 1875, when the rate of industrial growth was higher than ever before.

In the early 19th century mechanical steam powered devices began to replace the spinning wheel and handlooms and this had a remarkable impact on the British cotton and textile Industry. Almost simultaneously, a remarkable impetus for the industrial growth was provided by the railways, which began operating in England by about 1830. While until around 1830 the industrial revolution centered around England, the rest of Europe and America soon caught up with this momentum and in some sphere even overtook England. India caught up this momentum during the late 19th century, when Jumshedji Tata started Tata Iron and Steel Company (TISCO), and then on, the Indian cottage industry atmosphere changed to small and medium scale industry atmosphere.

Work Organization and Methods Engineering for Productivity.
DOI: https://doi.org/10.1016/B978-0-12-819956-5.00002-9

2.3 Evolution of management thinking

A significant contribution of the industrial revolution is the large scale industrialization and the outcome of factories. While this undoubtedly paved the way for overall development and better standards of living, it created new management problems, since the number of persons supervised by a single person and the levels of supervision have considerably increased. This initiated management thinking.

The growth of management thinking can be split into three phases.

- The first phase covering the late 17th to mid 19th century, mostly centered in Europe.
- The second phase from mid 19th century to mid 20th century. USA contributed considerably to this thought; Frederick Winslow Taylor's Scientific Management thought was the significant outcome of this phase. In fact, it is during this period that work study really developed.
- During the third phase after mid 20th century, specifically after 1960, considerable changes in management thinking have taken place, where in Taylor's scientific management thinking of job specialization has given way to modern thinking of total involvement of all employees in all the related function.

Thus the concepts of Total Quality Management, Total Productive maintenance, decentralization of routine function etc. have emerged. The customer and his views are today given the maximum weightage and quality is given the topmost priority.

2.4 Early pioneers in management thinking - Pre-F.W. Taylor period

In the chronological order, the following can be acknowledged as some of the early pioneers in management thinking, though there might be a host of others not cited here.

Richard Arkwright (1732–92), realized the value of job training to the operators in operating a new spinning jenny. He also developed a code for factory discipline.

J.R. Perronet of France, tried systematic overall timing of the manufacture of pins and arrived at a standard production rate as early as 1760.

Mathew Boulton started a factory in 1762, that was full of mechanical inventions and has a body of highly skilled craftsmen, His son Mathew Robertson Boulton, in conjunction with James Watt (Jr), preplanned and expanded this factory into a closely integrated modern engineering plant over 2 centuries ago. As early as in 1770, he Arranged weekly meetings of his partners and managers to examine business orders, pricing or any other issues related to production.

Robert Owen (1771–1858), worked for improved labor relations and employment conditions and was largely responsible Working hours for labor, Which later led to the introduction of the factory act of 1891.

Charles Babbage (1792–1871), recognized the necessity for detailed costs and invented the earliest calculating machine. In his book 'The Economy and Machinery and Manufacture' in 1832, he reflected the need for re-establishing general principles on the management business undertakings. He was aware of some of the dangers of time study and printed his own blank formats for the collection of data in main investigations.

Henri Fayol (1841–1925), was responsible for the concept of the principles of organization as well as the principles and functions of management.

Frederick Winslow Taylor (1856–1915), the most popular management expert of the 19th century dubbed as the father of Scientific Management. A special tribute is hence paid to him in the next paragraph.

2.5 Early pioneers in the development of management thinking in the 19th century

In the late 19th century, the scene of management thinking shifted to be centered primarily in United States, where a body of management thinking emerged as a result of the efforts of a number of individuals, primarily by Taylor and later strengthened by Gantt, Gilbreth etc. This new management thinking has come to be known as scientific management, a term coined by Taylor. It is during this period that work study had its strong foundations in the development of its concepts and techniques that are followed even today. The following pioneers can be cited as the milestones in the development of modern work study.

- *Frederick Winslow Taylor (1856–1915)*, who is undoubtedly a significant figure in the history and development of management thought and is generally acknowledged as the founder of the scientific management. He stressed the maximizing of production by following the three basic principles.
    - A defined task, determined by the definition of the job leading to the best operation sequence.
    - A definite time, established by stopwatch time study or estimated from standard data, and
    - A definite method developed by detailed analysis and recorded in the instruction charts.

Thus Taylor's significant contribution to work study is the timing of every operation and arriving at 'one best way' of doing a job. He also ventured to oppose the traditional line type of organization and stressed that every supervisor must specialize in a specific function that being 'a jack of all trades but master of none'. He has reflected all these principles in his book 'Shop Management' (1903).

Henri Gantt is credited with his famous Gantt charts, one of the most useful tools of production control even today. His name is also associated with the humanistic approach in management and wage incentives. He is credited to have developed the bonus plan, the forerunner to increase productivity in the later years.

Frank Gilbreth (1868–1924), initiated the idea of motion study and contributed a lot to bring to the present state of perfection. The 17 fundamental motions like grasp and search are called Therbligs, an anagram of his name. His wife ***Lillian Gilbreth***, a psychologist, whose name is always associated with him in the development of therbligs, went on developing more therbligs and adapting them to the home and similar environments.

Charles Bedaux, known for the concept of rating in work measurement. His Bedaux System for the measurement and remuneration of human labor was a popular tool as a productivity strategy amongst the world's business leaders.

2.6 Concepts of scientific management

Management is chiefly concerned with men. It is defined as the art and science of directing, coordinating and controlling human effort, so that the established objectives of an enterprise can be achieved in accordance with the established policies and procedures.

Scientific methods mean the systematic and logical procedure of observing and ascertaining the facts followed by reasoning and critical analysis conducted in such a way that the same can be fully utilized in building up a system of cooperation founded on the agreement, justice and integration of ideas.

We can hence define scientific management as the application of systematic and logical procedures in the ascertainment of the facts, reasoning and critical analysis in finding rational solutions, in coordinating and controlling human efforts and resources so that the established objectives of an enterprise can be achieved in accordance with the established policies and procedures.

2.7 Specific aims of scientific management

1. To identify waste and its causes, which may be in the form of inappropriate materials, manpower, machinery, manufacturing process or of capital. These may either be due to lack of coordinating efforts or unhealthy relations between employees and employers.
2. To eliminate waste after ascertaining the reason for it.
3. To unify the larger interests of labor, management and ownership and also to lower the costs by systematizing each and every process of the production, distribution and administration.
4. To unite higher wages with reduced labor cost and thus to ensure benefit both to the employee and employer.
5. To ensure a higher standard of living for the worker in proportion to his efficiency.
6. To increase the purchasing power of the consumer by lowering the selling price, thereby benefiting the community at large. It is thus important in social and economic movement.
7. To study the level of employment of labor and capital by gauging the industrial and market fluctuations.

2.8 Advantages of scientific management

A. Advantages to the employer

- (i) Scientific management lowers the cost of production.
- (ii) It minimizes labor troubles, strikes and lockouts.
- (iii) It guarantees prompt deliveries of goods.
- (iv) It increases production with marginal increase in capital investment.
- (v) It improves the quality of products through better and more efficient supervision.
- (vi) It hence increases the demand for the product.

B. Advantages to the workforce

(i) The workforce gets higher reward for their higher performance.
(ii) They work for shorter time, but still get higher output.
(iii) Working conditions are greatly improved and the workforce feels they are well looked after.
(iv) The system ensures steady employment, because the business is steady.
(v) The work fatigue is reduced.
(vi) Due to increased wages the standard of living increases.

C. Advantages to the community

(i) The employees, who constitute a greater part of the community, get direct benefits.
(ii) The system makes the community wealthier, healthier and more contented.
(iii) Absence of strikes and lockouts result in greater industrial stability.
(iv) Export trade increases and the inflow of money from abroad enhance the general prosperity.
(v) Periodical trade depressions are reduced to a minimum.

2.9 Misconceptions of scientific management

Unfortunately the term "Scientific Management" soon developed in a matter of another dispute. This term, coined by Taylor, implies a precision which cannot always be present when dealing with humans. This allowed a variety of interpretations wherein the chief principle of Taylor was distorted by unscrupulous managers wanting to ride over the workers in the name of scientific management. All essential requirements such as work load fixation, production targets and pay earned were determined scientifically by time study and arithmetic calculations. They felt the workers have no cause for complaint under this system of control, and the latter was ultimately treated just as one among other factors of production. Such a situation gave rise to the following misconceptions of scientific management.

1. It was felt that scientific methods have been adopted to promote exploitation of workers as explained above. That is, they are applied only to techniques like work study and the problems of industrial relations were overlooked.
2. Scientific techniques were presumed to be consisting only of long mathematical formulae which can be applied to all situations. Nevertheless, the first step, that is, the scientific method of observation of factors and their reasoning, is one which greatly influences the logical solution that can be given for a given situation.
3. The initial concept of scientific management presumes that the detail with which the problem is analyzed is identical to the whole. In fact the

whole is not only the sum of the parts, but also interrelationships, plus the individuality of the whole. For example, in Time study, the worker's output does not only depend on the operator's physical movements, but also on his aptitude, his relations with the fellow employees, etc.

4. The misgiving that the scientific methods can be applied to only measurable characters resulted in the exaggerated and the overstretched development of micro motion principles, ignoring altogether the psychological factor.
5. One immediate evil result of this concept is to chase the worker for higher productivity, with no or little improvement in factory layouts, working conditions, fatigue, etc.
6. Taylor's emphasis on extreme job specialization has subsequently been felt unrealistic and created new psychological problems.

2.10 Resistance to scientific management

Despite the above mentioned advantages, as explained in paragraph 2.7, scientific management soon developed in a matter of disrepute. Taylor's emphasis of precision in measuring work (time study) cannot always be present when dealing with humans. This allowed a variety of interpretations wherein the chief principle of Taylor was distorted by unscrupulous managers wishing to ride over the workers in the name of scientific management. Naturally, these so called efficiency experts, who had little knowledge of human relations met with strong opposition from labor. The disturbances spread to such an extent that the interstate Commerce Commission of the United States investigated and reported against time study in 1910, resulting in a Government ban in 1913 on time study in its undertakings for some period. This opposition to scientific management is similar to that against changes in operational methods as dealt in more detail in Chapter 22.

2.11 Birth of industrial engineers

Although scientific management as a distinct theory or school of thought was obsolete by the 1960s, the resistance to it being illustrated in the previous paragraph, most of its themes including analysis, synthesis, logic, rationality, work efficiency and elimination of waste, standardization of best practices, etc. are still important elements of industrial engineering and management today.

The later part of the First World War saw unprecedented demand for higher production in all factories, especially in the ammunition factories. This could have been well achieved by the target fixation and time study, but for the stiff opposition. Hence a need arose for specialists in work study with a good appreciation and understanding of human relations, who can also concentrate more on the management problems than the day to day technical and shop problems. This need gave birth to a new field of management specialists in industry with engineering background. This field was attributed the term Industrial Engineering, merely as an acknowledgment of the awareness of its importance in the industry.

Thus work study, which is the first basic function of the industrial engineers, continues to be their major function even today, even though the spheres of activities of the latter have broadened and embraced many new activities and applications whenever systematic investigation, analysis and planning is required, not only in industry but in other spheres like hospitals, libraries etc. Of late industrial engineering has spread its tentacles far into Operations Research, Computer Applications, Business Process Reengineering, etc. Fig. 2.1 depicts the above detailed evaluation process.

2.12 Industrial engineering and operations research

Industrial engineering develops safer and more productive methods of operations by systematic application of creativity. Operational research uses analytical skills in creating and using mathematical and analytical models to develop the shortest travel path and the quickest elemental time to perform an operation thereby developing optimal systems and operational procedures.

While Industrial Engineers use simulation, decision making tools, graph theoretical algorithms, optimization techniques and software to solve problems, Operations Researchers work on the science behind these tools, they develop new algorithms, design new optimization methods, design tools for analyzing systems.

In a nutshell, Industrial Engineering & Operations Research combine the two disciplines which focus on the operation of complex systems. Today, Industrial Engineers, apart from the traditional work study techniques, apply operations research in areas like manufacturing, logistics, production planning, facility design, service systems, etc. To aid the management to take decisions in complex situations. We can say that Industrial Engineering is the "engineering" while Operations Research is the "science" behind it.

Perhaps it is with this objective, several American Universities like MIT and Berkeley, emphasize Operations Research in the curriculum of Industrial Engineering.

2.13 Definition of work study

In the previous two paragraphs and Fig. 2.1, we have seen how industrial engineers are born with work study as their main function. In the following paragraphs, we shall study in detail what work study is, what are its major fields of activity,

In the first chapter, we have understood the basic concepts of work study and understood that work-study is a system of assessing methods of the most

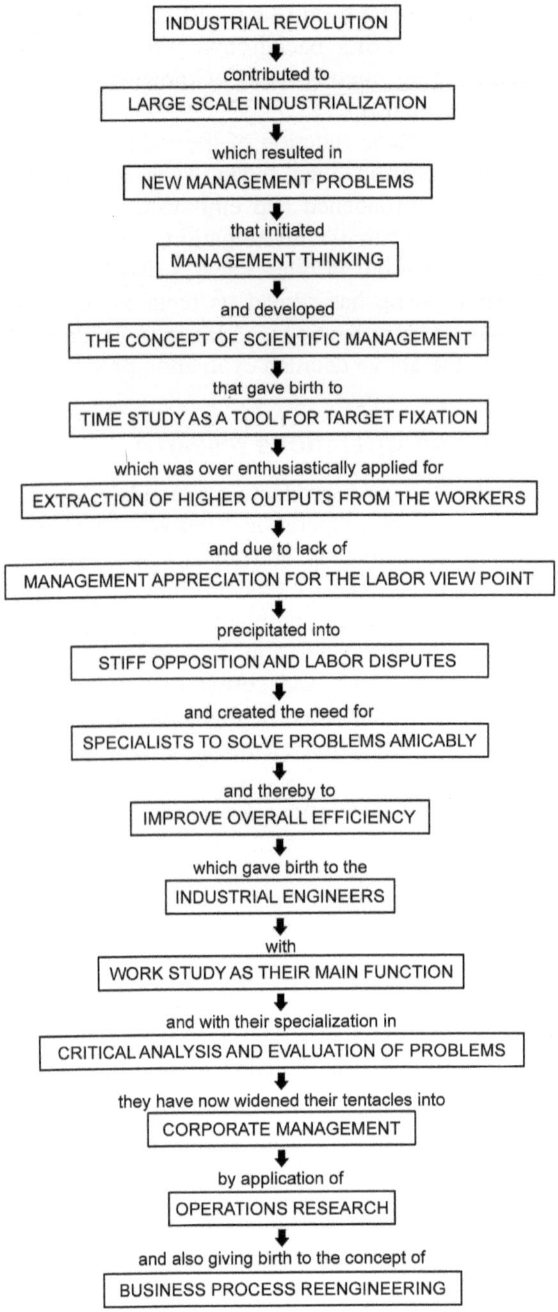

FIGURE 2.1 The genesis of industrial engineers.

efficient methods of working so as to achieve the maximum output and efficiency. For this objective, we need to know, the method of performing each operation and how long each operation takes. These two objectives are provided by the two wings of work study, via method study and work measurement.

From the definition of work study per paragraph 1.2 in Chapter 1, we see that work study has basically enveloped two components, the methodical study, and the work measurement. In this chapter, we will try to explain each of these components and see how they are related to each other and how they together contribute significantly to work study before we go into their more detailed precepts and practices in later chapters.

2.14 Earlier form of work study

Work study in earlier days was known as Time and Motion study, implying the coverage of time study to which Taylor is generally accredited. These two for long have been identified as the combined techniques of time and motion study. Time study refers to the timing of the operations with the use of a stopwatch, basically to set up the standard time of operation. Motion study refers to the study of the body motion in and around the workplace with a view to simplify the body motions or the operational procedures.

2.15 Work study vs. time and motion study

A comparison between the earlier and modern definitions of work study reveals the following points of discussion.

(a) The term time study is now replaced by the term work measurement that covers not only the stopwatch time study, but also the latest developments like Predetermined motion Time Systems.
(b) The term method study covers all the technique applied to critically examine the complete method of operation in order to eliminate the ineffective elements, whatever may be the situation or type of operation. As we will see in the next chapter motion study is restricted to the analysis of the finer body motions.
(c) In the early days, more emphasis was paid to time study rather than to motion study (now method study). That is more importance was attached to use work study of setting up production targets rather than in improving work methods.

Of the above, the last is of particular significance. With all his good intentions, Taylor paid undue emphasis on timing the operation and setting up standards. This, as we already saw earlier on this chapter, resulted in stiff opposition and forced later pioneers to review the concepts of time and motion study. In fact, Ralph Barnes named his famous book in 1937, as "Motion and Time Study", emphasizing that the search for proper methods should always precede the setting up of time standards.

Let us first look into the definitions of Time and Motion Study, and then cite the definitions in its two components, viz, method study and work measurement.

2.16 Definitions of time and motion study

We can cite some definitions for Time and motion study as follows

Time and motion study is a business efficiency technique combining the Time Study work of Frederick Winslow Taylor with the Motion Study work of Frank and Lillian Gilbreth (the same couple as is best known through the biographical 1950 film and book Cheaper by the Dozen).

Wikipedia.

Time and motion study, or motion and time study, is a basic set of tools used by industrial engineers to increase operational efficiency through work simplification and the setting of standards, usually in combination with a wage-incentive system designed to increase worker motivation.

Encyclopedia.com.

Time and motion study envelops the method for establishing employee productivity standards in which (1) a complex task is broken into small, simple steps, (2) the sequence of movements taken by the employee in performing those steps is carefully observed to detect and eliminate redundant or wasteful motion, and (3) precise time taken for each correct movement is measured

Business dictionary.

2.17 Broadened concept of time and motion study

Nevertheless, with the development of new techniques in these fields and their application to a wider range of activities, it was felt that the terms time study and motion study were too narrow and insufficiently descriptive. Hence, they are replaced by the modern terms of method study and work measurement, each of which are defined as below

We can cite some definitions for methodical study as follows

Method study is the systematic recording and critical examination of the existing and proposed ways of doing a work, as a means of developing and applying easier and more effective methods of operations, thereby reducing costs.

I.L.O.

Method study is subjecting each part of a given piece of work to close analysis to eliminate every unnecessary element or operation, as a means of approaching the quickest and best method of performing the work. It also includes formulation of incentive schemes, and improvement and standardization of equipment, methods, operator training, working conditions, etc. Also called methods engineering.

Business dictionary.

Method study is the process of subjecting work to systematic, critical scrutiny to make it more effective and/or more efficient. It is one of the keys to achieving productivity improvement.

Institute of Management Services.

Likewise, the definitions for Work Measurement can be cited as

Work measurement is the application of the techniques designed to establish the time for a qualified worker to carry out a specified job at a desired level of performance.

Wikipedia.

Work measurement is the determination of the standard time that a qualified worker should take to perform the operation when working at a normal place.

Institute of Management Services.

Work measurement is the application of time and motion study and activity sampling techniques to determine the time for a qualified worker to complete a specific job at a defined level of performance.

Business dictionary.

Work Measurement is a term which covers several different ways of finding out how long a job or part of a job should take to complete. It can be defined as the systematic determination, through the use of various techniques, by the amount of effective physical and mental work in terms of work units in a specified task.

http://www.managers-net.com.

The above defined two wings of work study can be represented as in Fig. 2.2.

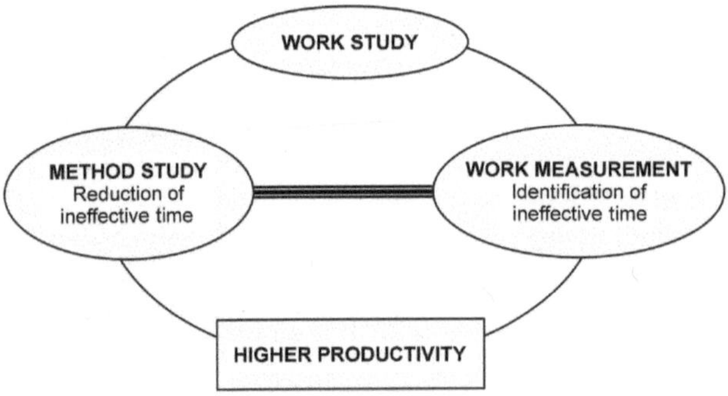

FIGURE 2.2 Method study and work measurement.

2.18 Relationship between method study and work measurement

Though method study and work measurement adapt different procedures, basically they are linked closely. Method study is involved in the reduction of the work content of a job, while work measurement is involved with the investigation and identification of the ineffective time associated with the job, and the subsequent establishment of the time standards for the improved operation. The relationship is depicted by Fig. 2.3, which is an extension of Fig. 2.2, but indicates the several stages of each.

Even though for any method study project work measurement is necessary to determine the best method, it is generally advised that preliminary method study, at least the first 3 steps of select, record and examine, to assess if the method demands a change, should precede work measurement. Obviously, there is no point in timing an operation which in any case would be improved and become obsolete immediately after the timing. Nevertheless, the existing method may be timed for comparing the existing and proposed methods of operations.

In most circumstances where the work study project is intended for job simplification or improvement, this advice holds good. Despite the discussions in paragraph 2.9 of this chapter, there are certain circumstances, where preliminary work measurement must be carried out in order to achieve the following.

(a) Discover and eliminate lost or ineffective time.
(b) Arrive quickly at the work values, which will facilitate bonus payments.
(c) Provide a measure of work content of jobs for comparison purposes.
(d) Provide a quantitative means of assessing the later productivity improvements brought out by subsequent method study projects, and
(e) Provide a preliminary basis for the allotment of priorities or determination of areas of study in the proposed method study projects.

An example where work measurement proceeds method study is work sampling, where it is necessary to take preliminary time studies and assess

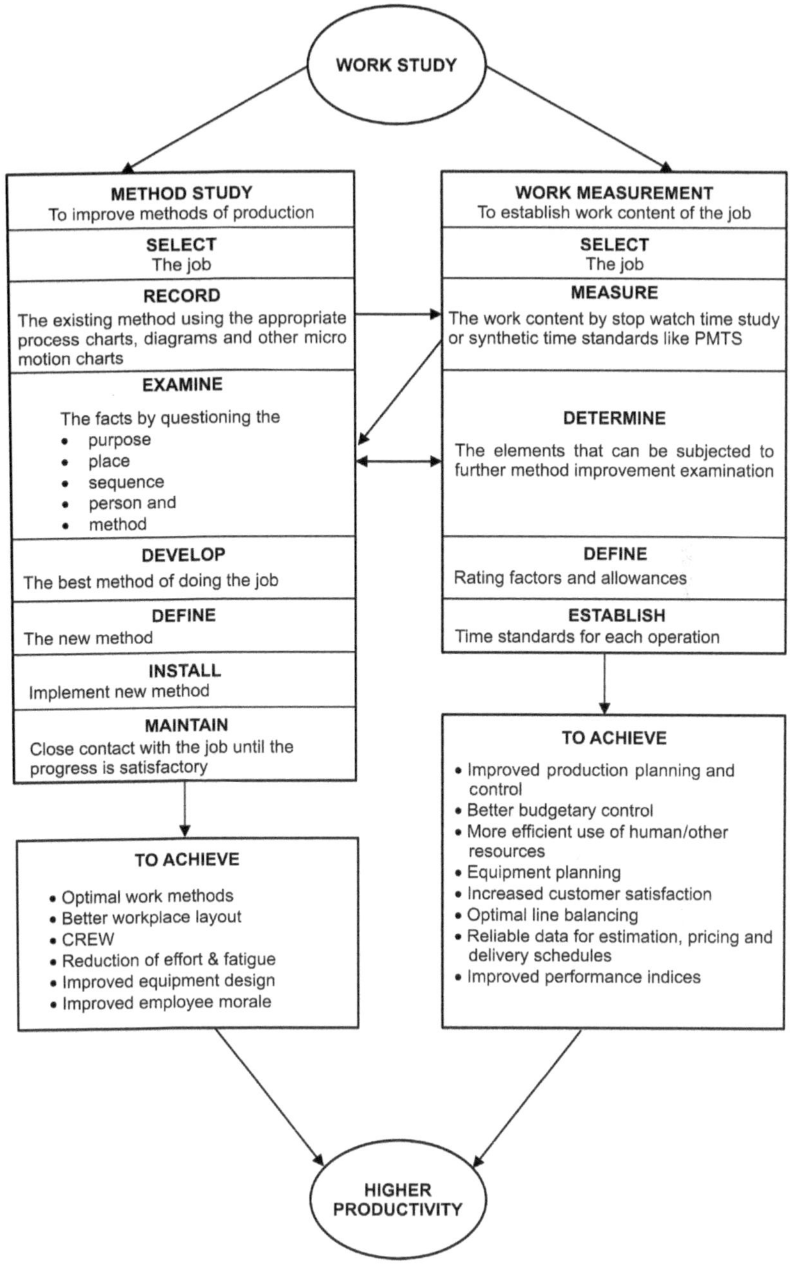

FIGURE 2.3 Method study vs. work measurement.

why and by how much the ineffective time occurs, so that management can take action to reduce it. This is further detailed in Chapter 17 on Work Sampling.

2.19 Conclusion

Though method study and work measurement have different procedural concepts and make use of different techniques, they are quite dependent on each other towards achievement of the productivity goal, and are generally clubbed together. It may be noted that work measurement should be used as a supplement to method study for determining the time saved by the improved methods. It should never be used for the sole purpose of setting up work standards or operative targets.

Criteria questions

1. Discuss the evolution of scientific management concept. (2.2)
2. Illustrate the contribution made by Frederick Winslow Taylor to work study. (2.5)
3. Justify the statement the history and development of work-study is closely related to the history and development of scientific management. (2.7)
4. What in your opinion is the reason for the workers opposition to time study? (2.9)
5. Trace the evolution of work study engineers from the days of industrial revolution. (2.11)
6. How operations research is related to industrial engineering? (2.12)
7. Distinguish between method study and time study. (2.18)

Further reading

1. Batty J. *Industrial administration and management.* 3rd ed. ELBS; 1974.
2. Hammond RH. History and development of industrial engineering. In: Carson G, editor. *Production handbook*, 5th ed. Arnold Press; 2011.
3. Urwick Col. LF. Development of industrial engineering. In: Maynard HB, editor. *Industrial engineering handbook*, 2nd ed. McGraw Hill; 1963.
4. Kiran DR. *Elements of production planning and control.* BS Publications; 2017.
5. Tarachand, *Engineering economics.* 2nd ed., vol. II. Nemchand Publ; 2012.
6. Kart, Rosenzwig. *Organisation and management: a systems approach.* McGraw Hill; 1970.
7. Spreigel, Lansburgh, *Industrial management.* John Wiley & Sons; 1966
8. Barnes RM. *Motion & time study.* 7th ed. John Wiley & Sons; 1980.

Chapter 3

The concepts of productivity

3.1 Introduction

Productivity is an age old axiom for the measure of the efficiency of a person, machine, factory, system, etc., in converting inputs into useful outputs. Since work study generally is referred to as the tool to increase productivity, it is imperative to understand more of productivity in industrial and other situations before our going into the details of work study. This chapter as well as the next one, hence attempt to create an appreciation of the role played by work study in achieving the goal of increased productivity.

As the topic suggests, modernization and innovation or the development of new products and processes may be some of ways to increase productivity, but the developing countries with their difficult foreign exchange situation cannot afford to think of modernizing the industrial system, by importing know how for the new products or processes. In short, we cannot afford that type of innovation that necessitates larger inflow of inputs. What is needed today is that aspect of innovation, that improves the functional design of a product or simplifies the method of operation, so that with the least increase in inputs, the output is considerably raised, thereby increasing the operational efficiency. This point can be better understood by Fig. 3.1.

This exactly what work study does. As we have seen in the previous chapter the term work study is not new to industry sector. One of its wings, viz., time study has long been used as a tool to set up targets, thereby maintaining productivity. But this aspect has been overemphasized in the industry, and as a result the impression we get of work study has been a function of the stopwatch time study man, who is generally eyed with suspicion by the labor as a management agent to set targets and reduce their wages, this has resulted in a complete lack of appreciation of the benefits of the other wing of work study, viz., method study or work simplification, which by means of its systematic investigation and critical examination techniques, aims at increasing productivity with little inflow of inputs. This fact that the continuous improvement or what the Japanese call as Kaizen, can achieve higher cost reduction than what could be achieved by expensive innovations, is further discussed in Chapter 11 on kaizen.

As an attempt to promote the full-fledged practice of work study, the following paragraphs present an anatomy of productivity, highlighting work

Work Organization and Methods Engineering for Productivity.
DOI: https://doi.org/10.1016/B978-0-12-819956-5.00003-0

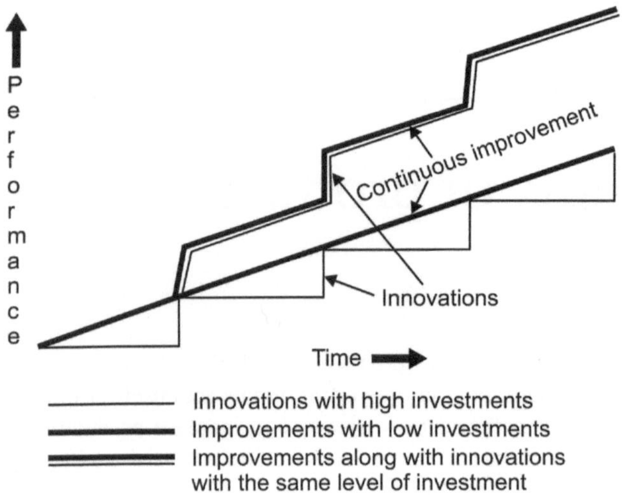

FIGURE 3.1 Improvement vs. innovation.

$$\text{PRODUCTIVITY} = \frac{\text{OUTPUT}}{\text{INPUT}}$$

FIGURE 3.2 The productivity formula.

study as the most significant and practical management tool available to the industrial manager in achieving higher productivity.

3.2 The concept of productivity

Productivity is related to the goods and services produced in relation to the resources utilized in producing the same. In a broad sense, it is the ratio between the output of a certain process and the input of the resources employed. We can express this in a simple formula Fig. 3.2.

3.3 Some definitions of productivity

I.L.O. in one of its publications, *Higher Productivity in the manufacturing industries* defined Productivity as the ratio between the output of wealth and wealth input of resources used in the process of production.

Some of other more specific explanations offered on productivity are:

> *Productivity is an attitude of mind; it is the mentality of the progress of the constant improvement of that which exists. It is the certainty of being able to do things better today than yesterday and continuously. It is the constant adaptation of economic and social life to the changing conditions. It is the continual*

effort to apply new techniques and methods. It is the faith in the human progress.

European Productivity Agency (EPA).

In the widest sense, it may be said that productivity is the measurement of the economic standards of the means.

The organization for European Economic Cooperation (OEEC).

Productivity is a measure of the efficiency of a person, machine, factory, system, etc., in converting inputs into useful outputs. Productivity is a critical determinant of cost efficiency.

Business Dictionary.

Productivity is an average measure of the efficiency of production expressed as the ratio of output to inputs used in the production process.

Wikipedia.

Productivity is an economic measure of output per unit of input. Inputs include labor and capital, while output is typically measured in revenues and other GDP components such as business inventories.

Investopedia.

Productivity is the rate per unit area or per unit volume at which biomass consumed as food by other organisms is made by producers.

Webster Dictionary.

Productivity is the rate at which a company or country makes goods, usually judged in connection with the number of people and the amount of materials necessary to produce the goods.

Cambridge Dictionary.

In spite of providing such endless chain of explanations and explanatory definitions for productivity, the simplest definition is still as given earlier, viz., the ratio of output to input.

In other words, productivity would mean

for the nation	The output of goods and services needed by the community by optimal use of resources like materials, equipment and manpower available.

for the industry	The output of products with the optimal use of the capital assets, materials and services of employees.
for individuals	Saleable value of the work achieved in return for the use of individual resources and those provided for personal use.

Even though productivity is referred to as the input-output ratio in the above definitions, it is in effect the process of harnessing the capacity to increase the outputs by optimal utilization of all the resources employed at the most economical cost of production.

3.4 Productivity vs. production

We should not confuse productivity with production, which is merely the quantum of output irrespective of the resources used up. For example, a certain forge shop has 10 forging machines, if the output of each machine is 100 pieces per shift, then the total production or output for the shop is 1000 pieces per shift. In general, the words output and production are synonymous.

We can increase the production by merely increasing the resources without consideration to the cost. For example, the production of the forging shop cited above can be increased from 1000 pieces per shift to 1500 per shift, by merely increasing the number of machines from 10 to 15. But this increase in production need not necessarily increase the productivity, which is the output for each input resource that is the output of each machine, which continues to remain the same at 100 pieces per machine per shift. The distinction is more explained by the following illustrations.

Example: If 100 men of group A can lay 5 km of road per week, and if 150 men of group B can lay 6 km similar road per week, whose productivity is higher?
Productivity of group A = 5 km/100 = 50 m per man per week.
Productivity of group B = 6 km/150 = 40 m per man per week.
Hence the productivity of group A is higher than that of group B
Example: The total production per shift of group A of 20 wood screw making machines is 6000 gross, while the production of group B of similar machines numbering 30, is 7500 gross. Whose productivity is higher?
Productivity of group A = 6000/20 = 300 gross per shift per machine.
Productivity of group B = 7500/30 = 250 gross per shift per machine.
Hence, we can say that the productivity of the first group of machines is thus higher than that of the second group.

3.5 The input-output concept

This concept of productivity as the ratio of the output of the inputs, is applicable in any situation, involving a conversion process, whether it is a manufacturing industry or agriculture or trading or even an educational institution. In view of its significance, this concept is dealt more in detail in Chapter 5.

3.6 Connotations of productivity

Apart from the above meaning of productivity, we can also connote the following terms related to productivity

- Physical Productivity, which is the ratio of the quantum produced with the resources consumed (usually labor effort in terms of man-hours, days, or months).
- Functional Productivity, which is a ratio of the amount of the functionality delivered or value added to the resources or effort consumed. Functionality may be measured by use cases, requirements, features, or functions, as further discussed in Chapter 18 on Value Analysis.
- Economic Productivity, which is a ratio of the value of the product produced for the cost of the resources used to produce it. It helps to evaluate the economic efficiency of an organization.

It may be noted that while understanding economic productivity is essential in making good decisions about outsourcing and subcontracting, it alone is not used to predict project costs since the outcome can be affected by many factors outside the control of the project, like sales volume, inflation, interest rates, substitutions in resources or materials etc., as well as all the other factors that affect physical and functional measures of productivity.

3.7 The measure of productivity

The measure of the term Productivity varies due to differences in technology, differences in the efficiency in the production process and the differences in the environment in which the production unit operates.

There are various productivity measures depending upon the purpose of productivity measurement and, in general, on the availability of the data. As explained below, the measures can also relate gross output to one or several inputs and those which use a value-added concept to capture movement of output.

We can classify these measures into 2 broad groups

- Single factor productivity measures (relating to measures of output to a single measure of input)
- Multifactor productivity measures (relating a measure of output to a bundle of inputs).

As explained above, the unit we used for measuring productivity as seen from earlier paragraphs, can either be in terms of output per man per unit time or output per machine per unit time or in other words, the output per man hour or machine hour. But in actual practice, there can be hundreds of such units in which productivity can be expressed, depending upon the situation. In such cases only a single input is considered. For example, in our previous example on the road laying in paragraph 3.4, we expressed the productivity in meters per man per shift, that is, we considered only the human input, whereas there could be other inputs in the form of material, tar, etc. Again in comparing the performances of different cars, we compare the fuel productivity or the fuel efficiency of the vehicle in km per liter (kpl), since fuel is the major input.

Obviously, to get an absolute measure of productivity, we need to express both the inputs and outputs in the same units. One of the most convenient ways of doing this is to assign a money value to each input and output. That is, productivity can be expressed as the ratio of the total value of the outputs of the total cost of the inputs as already expressed in Fig. 3.2.

Or

Where

Tp = Total Productivity
To = Total Output
Ti = Sum of all inputs

We can also estimate the productivity of each input separately and this is called the partial productivity.

3.8 Other measures of productivity

As illustrated in the earlier paragraphs, the following indices can be cited as the other partial measures of productivity based on individual inputs measured in money units.

$$\text{Labor Productivity} = \frac{\text{Total output}}{\text{Labor input}}$$

$$\text{Material Productivity} = \frac{\text{Total output}}{\text{Material input}}$$

$$\text{Energy Productivity} = \frac{\text{Total output}}{\text{Energy input}}$$

Total Productivity measure (TPM) or Total Factor Productivity is the ratio of the total tangible outputs of all products and services to the total tangible resource inputs

$$\text{TPM} = \frac{\text{Total tangible outputs}}{\text{Total tangible inputs}}$$

Where,

Total tangible output = Value of finished goods produced, value of partial goods produced + inventory + dividends from securities + interest + other income in money units and
Total tangible input = Value of labor + capital + material + energy and other expenses in money units.

Since the value of the money keeps varying year by year due to inflation, etc., the productivity is generally adjusted to the base year by a factor called the factor of deflation, which is expressed as the ratio of the current year price index to that of the base year.

$$\text{Factor of deflation} = \frac{\text{Current year price index}}{\text{Base year price index}}$$

3.9 Levels of productivity measurement

Depending upon the specific application of the productivity measurement, we can formulate these in 5 basic levels, as illustrated in the Table 3.1.

TABLE 3.1 Levels of productivity measurement.

Sl. no	Level	Measure	Application
1	International	Productivity indices from different countries	Compare the growth and competitive position among nations
2	National	Develop economic indicators	Enable the country to plan its resources on a national basis
3	Sectorial	Indices from different companies of this sector	1. Compare the performance of sector members for healthy competition 2. Plan their manpower requirement
4	Company	All productivity measures per para 3.7	Set up goals and plan the future requirement.
5	Individual	Develop performance measures	1. Compare index of each resource like performance and bonus earning 2. Self-improvement in these performances.

3.10 The concept of increase in productivity

Nevertheless, even the above expression invariably becomes difficult in many cases due to the fact that the inputs in all cases cannot be expressed in rupee value. This is more so in view of the complexity and variety of inputs and outputs.

Hence productivity is better understood by the concept of percentage increase in productivity. That is, we can compare the productivity of a certain resource or input before and after effecting, a change or improvement,

For example, if we can increase the produce of agricultural land from 1 tonne to 1.5 tonnes per acre by better farming methods like irrigation and fertilizers, without incurring much increase of capital expenditure, the productivity of the land is said to have increased by 50%.

Similarly, if an operator is able to increase his output from 100 pieces per Shift to 120 pieces by simplified methods, his productivity is said to have increased by 20%.

3.11 Factors that drive productivity growth

We can identify five Factors that Drive productivity growth as illustrated below

- Investment is in physical capital like machinery, equipment and buildings. It is believed that the availability of more capital enables production of more and better quality output.
- Innovation, metamorphosing of new ideas into new technologies, new products, etc. which can enable working faster and more efficiently to boost productivity.
- Skills, defined as the quantity and quality of labor needed to take advantage of the above investments and innovations.
- Enterprise, which enables seizing of new business opportunities by both start-ups and existing firms by new ideas and technologies.
- Competition improves productivity by creating incentives to innovate and ensures that resources are allocated to the most efficient firms. It also forces existing firms to organize work more effectively through imitations of organizational structures and technology.

3.12 How to increase productivity?

Returning to our basic formula on productivity viz, we can compare it with the simple arithmetic expression of

We know that A can be increased in any of the following manners:

(a) increase the numerator with no change in the denominator,
(b) decrease the denominator with no change in the numerator,
(c) increase the numerator substantially with marginal increase in the denominator,

(d) decrease the denominator substantially, with marginal reduction in the numerator, or
(e) increase the numerator, while decreasing the denominator.

We can adapt these principles to increase productivity by equating the numerator to output and the denominator to input.

1. ***Increasing the output for the same input:*** If the output of a machine is increased from 200 pieces per shift to 220 pieces per shift, without changing any input, the productivity is said to have increased by 10%.
2. ***Reducing the input for the same output:*** If the input of certain process is reduced by 20% without affecting the output, then the productivity is said to have increased by 25%.
3. ***Increasing the output with marginal increase in the input:*** In practice, this is the most popular method of increasing productivity. If the output of a certain operation is increased by 25% after providing a certain fixture which increases the machine cost by 10%, then the productivity is said to have increased by 13.63%.

   Fig. 3.3 below illustrates a case study for this increase in productivity. In a press shop, the weight of the steel strip required for the production

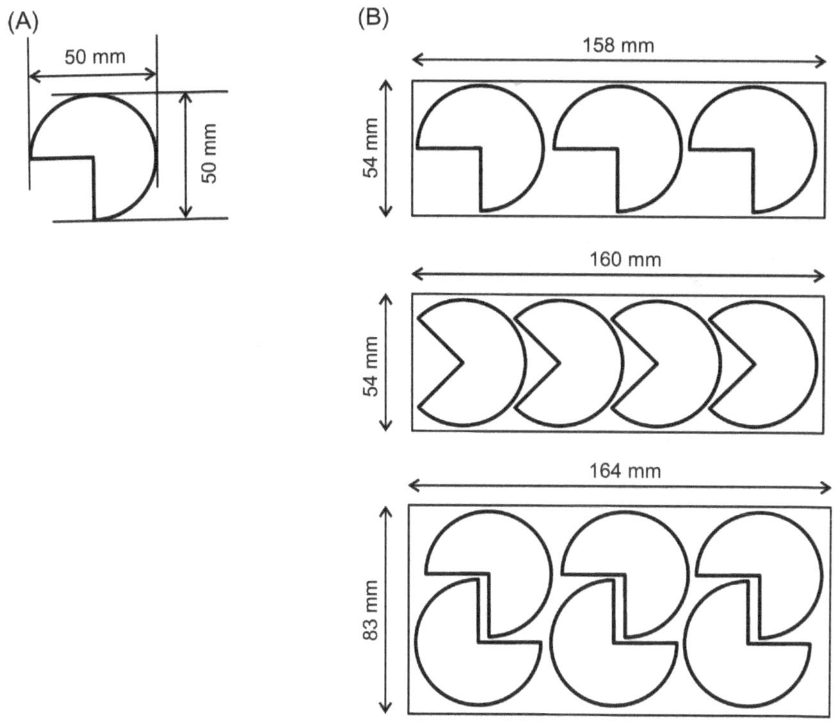

FIGURE 3.3 Illustration of reduction of input.

TABLE 3.2 Strip area for the production of the blank.

No.	Alternative	Feed (mm)	Strip width (mm)	Strip length per 1000 pcs (m)	Strip area per 1000 pcs (m^2)	Percentage reduction of scrap
1	A	52.0	54	158/ 3 = 52.7	2.848	–
2	B	37.8	54	160/ 4 = 40	2.160	24.1%
3	C	54.0	83	164/ 6 = 27.3	2.265	20.5%

of a certain blank, is reduced from 20 kg per 1000 blanks to 16 kg per 1000 blanks, as illustrated in Fig. 3.3. Also the output is doubled by increasing the die cost by 20% (Table 3.2).

In the third case, the productivity is said to have increased by

Example: The output of a certain drilling machine is 60 pieces per hour. By providing a certain jig, which has increased the input costs by 10%, the output for the same operation has increased to 90 pieces per hour. Calculate the increase in productivity.

Hence Percentage increase in productivity

4. ***Reducing the inputs considerably at a marginal loss of output:*** In a certain machine shop an effective preventive maintenance program has been introduced. By this, the total operation and maintenance cost of the machine shop is reduced by 20%. Nevertheless, this preventive maintenance program needed regular stoppages of the machine, reducing the total output by 5%. Here despite the reduction in output, the productivity is said to have increased by
5. ***Increasing the output with reduced inputs:*** This yields a two way benefit and is the most preferred form of increasing productivity, though it is less practical.

Example: On a certain machining operation, a methods program has been carried out, resulting in a 20% increase in the output of the operator. It also reduced the input costs by 5%. The productivity in this example is said to have increased by

3.13 Stewart's 12 step productivity improvement strategy

W.T. Stewart, in 1978, proposed a 12-step productivity improvement strategy for organizations based on a systems perspective. Though simple,

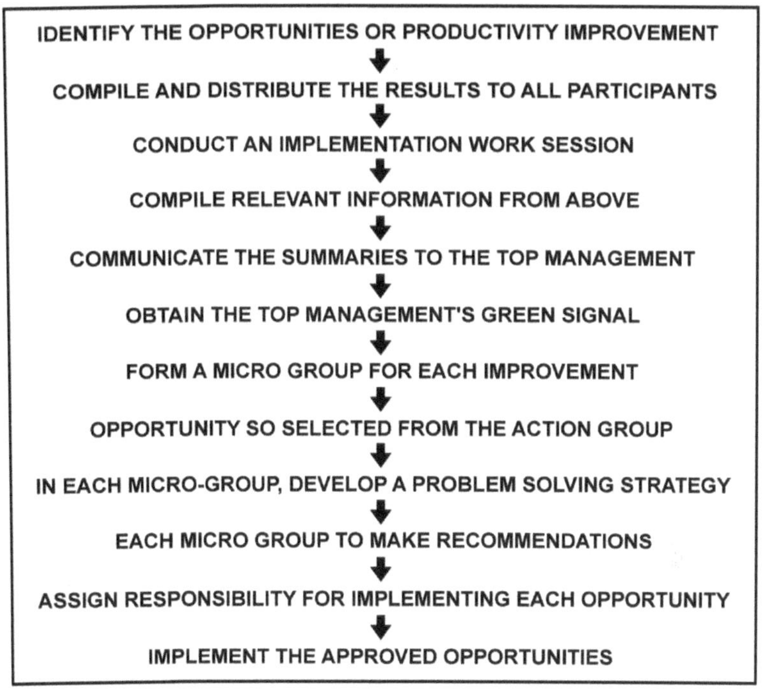

FIGURE 3.4 Stewart's productivity improvement strategy.

straight forward and common sense oriented procedure, this is applicable to any recommendation and has come to be known as Stewart's strategy, as illustrated in Fig. 3.4.

Subsequently, in 1979 Aggarwal and later other authors proposed the following modified 11-step procedure for productivity improvement, per Fig. 3.5.

3.14 Sumant et al.'s productivity improvement techniques

In 1990 David Sumant, Vincent Omachinu and Mario Beruvides of Miami University categorized the several techniques of productivity improvement into 5 basic groups as follows

1. **Technology based techniques**
   - (i) Computer Aided design (CAD)
   - (ii) Computer Aided manufacturing (CAM)
   - (iii) Integrated CAM
   - (iv) Robotics
   - (v) Laser Beam Technology
   - (vi) Group technology
   - (vii) Computer graphics
   - (viii) Emulation

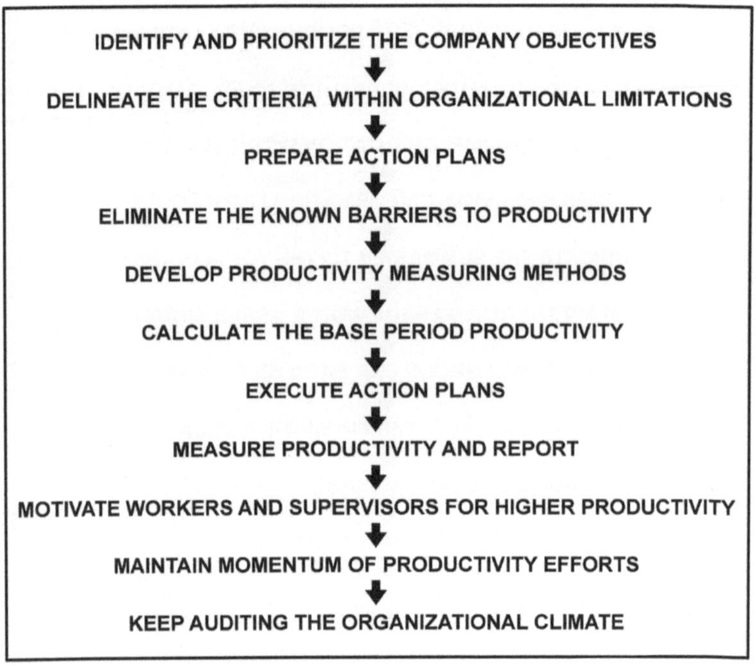

FIGURE 3.5 Modified productivity improvement strategy.

- **(ix)** Maintenance management
- **(x)** Energy conservation

2. Employee based techniques

- **(i)** Individual financial incentives
- **(ii)** Group financial incentives
- **(iii)** Fringe benefits
- **(iv)** Employee promotions
- **(v)** Rewarding
- **(vi)** Job enrichment
- **(vii)** Job enlargement
- **(viii)** Job rotation
- **(ix)** Worker participation
- **(x)** Skill enhancement
- **(xi)** Management by objectives
- **(xii)** Learning curve
- **(xiii)** Communications

3. Product based techniques

- **(i)** Value engineering
- **(ii)** Product diversification
- **(iii)** Product simplification
- **(iv)** Product standardization

(v) Research and development
(vi) Product reliability improvement
(vii) Advertising and promotion

4. **Task based techniques**
(i) Methods engineering
(ii) Work measurement
(iii) Job design
(iv) Job evaluation
(v) Job safety design
(vi) Ergonomics
(vii) Job hazard analysis
(viii) Production scheduling
(ix) Computer aided data processing

5. **Material based techniques**
(i) Inventory control
(ii) Material management
(iii) Material requirement planning
(iv) Quality control
(v) Material handling systems
(vi) Material reuse and planning
(vii) Material layout planning

In a nutshell, as suggested in Chapter 11 on Kaizen, the terms Kaizen, CREW et al can be cited as synonyms for method study except minor variations in conceptual application.

3.15 The benefits of higher productivity

A meeting convened by I.L.O. of the experts on Productivity in Manufacturing Industries in 1952, summed up the benefits of productivity as follows in its proceedings:

- Larger supplies of both consumer goods and of capital goods at lower costs and lower prices,
- Higher real earnings,
- Improvements in working and living conditions, including shorter working hours,
- In general the strengthening of the economic foundation of the human well being.

We can also cite the following other benefits by way of increased productivity.

A. **To the operatives**
- Higher productivity means higher wages, especially for the piece rated workers and higher bonus to all workers,

- More job security and job satisfaction,
- Boosted up morale and healthier industrial relations,
- Happier domestic life

B. To the company

- More profits due to reduced costs of production,
- Increased production volume and increased sales returns,
- Higher and continuous production, reducing the sales price and increasing better customer satisfaction,
- Better industrial relations and hence smoother management,
- Improvement in public image, resulting in higher benefits and name.

C. To the nation

- Higher profits earned by the firm would bring in higher tax revenues,
- Export trade is developed, bringing in more foreign exchange,
- Industrial relations would improve, providing healthy examples for other communities,
- Increased overall standard of living, and the consequent prosperity of the nation,
- Higher productivity in some companies induces the less productive companies to improve for their own survival.

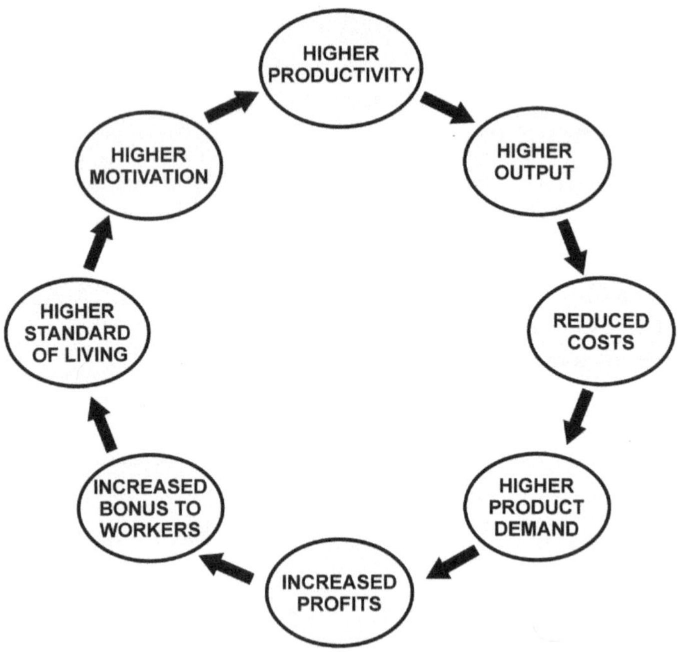

FIGURE 3.6 Chain reaction of increased productivity.

3.16 Productivity and standard of living

Since a country has limited resources, any increase in its productivity goes in a long way to help it to produce more goods thereby improving the general standards of living. Presuming that all other factors like inflation rate are in comparable terms, this statement can be explained by the Fig. 3.6, which depicts the chain reaction of higher productivity.

3.17 Conclusion

As rightly highlighted by I.L.O., higher productivity not only generates higher profitability to the organizations, it enables improvements in working and living conditions, including shorter working hours to the workforce, strengthening their economic well-being.

Criteria questions

1. Distinguish between innovations and improvements. (3.1)
2. How do you define and distinguish between Production and productivity? (3.3, 3.4)
3. Distinguish between Physical Productivity and Functional Productivity. (3.16)
4. Explain some of the measures of productivity. (3.8)
5. What is the meaning of Total tangible output? (3.8)
6. What factors drive productivity growth? (3.9)
7. Illustrate Increasing the output with marginal increase in the input. (3.12)
8. Discuss Stewart's 12 step productivity improvement strategy. (3.13)
9. Explain some of the benefits of Higher Productivity to the operatives. (3.15)
10. Justify the statement "Productivity and standard of living". (3.6)

Further reading

1. I.L.O. *Higher productivity in manufacturing industry*, I.L.O. 1972.
2. I.L.O. *Introduction to work study*. 3rd ed. I.L.O. 1979.
3. Schultz CL. National income analysis: a report. J Polit Econ 1964;460.
4. N.P.C. *Productivity and national growth*. Delhi: N.Po.C. 1972.
5. Faraday JE. *The management of productivity*. British Institute of Management; 1971.
6. Kiran DR. *Total quality management: an integrated approach*. USA: BS Publications/ Elsevier; 2016.
7. Kiran DR. *Production planning and control: a comprehensive approach*. BS Publications; 2017.
8. Kiran DR. *Productivity and work study in industry*. Tanzania: Work study Seminar; 1980.
9. Kiran DR. *Programmed instruction manual for NCES - Paper III-3, N.P.C.* 1985.
10. In: Carson, editor. *Production handbook*. Arnold Press; 1977.
11. IS 6363. Glossary of terms used in *work study*.

Chapter 4

Factors affecting productivity

4.1 Introduction

In the previous chapter, we made an attempt to get a concept of productivity and the principles behind the basic methods to increase productivity. In this chapter we shall analyze the various factors that contribute to the excess work content and the ineffective time added to the basic work content.

Symptoms of low productivity

- Congested space
- High rejects and returns
- Wasted resources
- Variable quality in products
- No accurate cost estimates
- No obvious work flow
- High inventory levels
- High level of overtime
- Missed deadlines
- Excessive equipment down time
- Inadequate overhead recovery
- Unachieved targets
- Unrecovered investment costs
- Inaccurate staffing levels
- Lack of appropriate productivity measures

4.2 How ineffective methods cause low productivity

As seen in Fig. 4.1 below, the ineffective time is present in all the elements of operations, but cannot be visualized normally, like an iceberg in the ocean. What you see during your patrolling around it, is only the tip of the iceberg of ineffective elements of operation. The major portion of the ineffective elements is hidden below the visual field, and some of the factors that cause this are as indicated in the figure, which are further explained in subsequent paragraphs. It is the systematic method study as detailed in the subsequent chapters, that would help us to identify and eliminate these elements.

Work Organization and Methods Engineering for Productivity.
DOI: https://doi.org/10.1016/B978-0-12-819956-5.00004-2

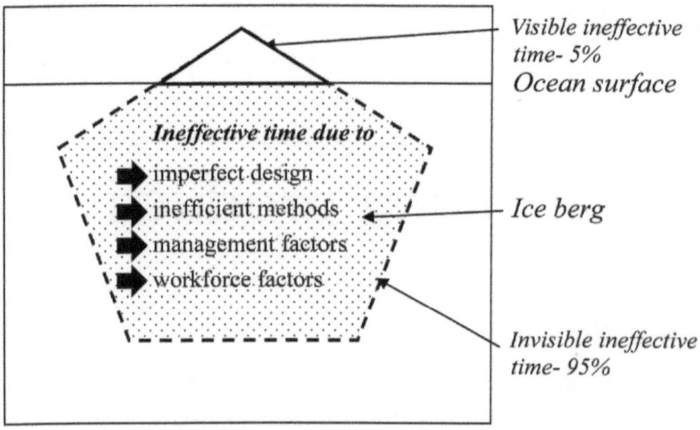

FIGURE 4.1 Visible and normally invisible ineffective operational Times.

4.3 Factors affecting the enterprise productivity

There can be three concepts of classifications of factors affecting the enterprise productivity.

- Internal and External factors, that control the flexibility of the change implementation as detailed in para 4.4.
- Hard and Soft factors propounded by Joseph Prokopenko in his book Productivity Management, as detailed in para 4.5.
- Factors contributing to ineffective time to the basic work content of a job, propounded by ILO in its book, *Introduction to Work Study* as detailed in paras 4.6 to 4.16.

4.4 Internal and external factors

The factors that influence the productivity of an organization, as illustrated in Fig. 4.2, can either be internal or external. Joseph Prokopenko, in his book *Productivity Management* edited by ILO (1987) classifies the factors that affect the productivity of an enterprise primarily into the internal and external factors. Let us distinguish between them so as to sort out our priorities and make effective productivity improvement.

Internal factors are those you can control many of them like the workplace layout and the functional design policies as well as the following,

- Human resources like employees, target audiences, and volunteers
- Access to natural resources, patents, copyrights, and trademarks
- Current processes like employee programs, software systems, and departmental hierarchies

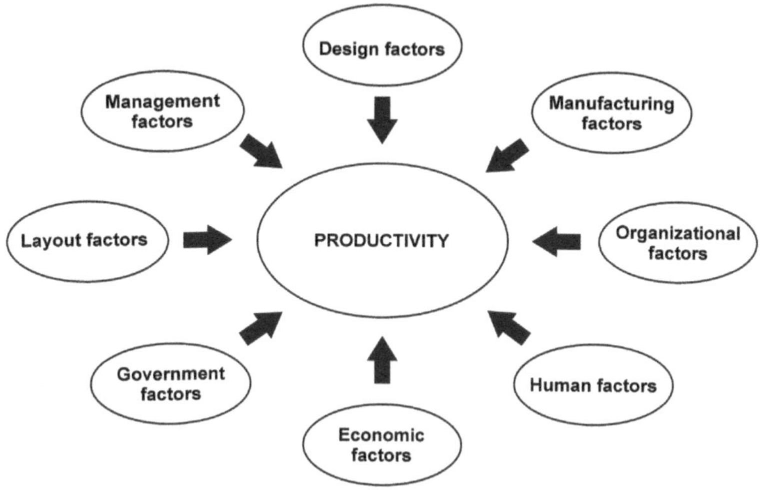

FIGURE 4.2 Factors affecting productivity.

Changing internal factors usually involves some indirect costs, such as lost productivity while direct costs include hiring of new employees.

External factors are all those things that are beyond our control and some of the external factors are those that affect the process of productivity improvement like

1. Customer's functional requirement of a product,
2. Unavailability of the optimal materials,
3. Competition and
4. Government regulations

Making a traditional list of pros and cons or conducting a simple return-on-investment analysis can add clarity to virtually any productivity improvement decision.

4.5 Hard and soft factors

Joseph Prokopenko, further classifies the Internal factors as

- Hard factors, that are well established and are more difficult to be changed like product design, plant & equipment and technology. They influence business practices in a direct and fundamental way.
- Soft factors that can be changed, if not easily, at least by persuasion, like personnel, organization & systems, work methods and management styles. They revolve around human beings, and hence are flexible, unpredictable, not easily calculable and not easy to grasp. On the lighter side, we can say that, in a way soft factors are harder to manage than the hard factors.

TABLE 4.1 Comparison between hard and soft factors

Sl. no.	Hard factors	Soft factors
1	Measurable. We can identify, collect, measure, almost grasp them.	Unmeasurable. They are flexible, differently interpretable, unpredictable, not easily calculable and hence difficult to grasp.
2	Goals and objectives bare clear and fixed.	Goals are flexible subject to need negotiations
3	They influence business practices in a direct and fundamental way	Management concepts change on a regular basis. Change is constant.
4	Power is clear, known to all	Power diffuses and frequently unknown
5	Unidirectional view of organization	Conesus view of organization
6	Evolved during W Taylor days to meet the needs of modern technical and industrial systems	Evolved during 1960 s in response to modern management thinking, considering difficulties in using hard factor approach
7	Aims to solve problems	Aims to appreciate and develop human oriented solutions to solve problems
8	Organizations as profit entities	Organizations as social entities
9	Analyst detached from solution	Analyst part of the solution
10	Examples are process approach, system approach and factual approach	Examples are customer focus, leadership, involvement of people, continuous improvement and mutually beneficial supplier partnership.

It may nevertheless be noted here that even the hard factors, are not rigid, but the systematic & critical analysis, integrated approach, etc., like value analysis, work study and other optimization studies would help in reducing the negative impact of these hard factors thereby resulting in higher productivity. Table 4.1 compares these factors.

4.6 Factors effecting the productivity vs. the basic work content

The following paragraphs as visioned by ILO in its book, *Introduction to Work Study* explain the reclassification of these factors based on four types of ineffective time caused by them on the basic work content of producing a component.

4.7 The concept of the work content of a product or an operation

Every product or component has an inherent work content associated with it, which is necessary to produce the same from the raw material stage. The work content, as the word implies, is the work contained in the product measured in time units, say in man minutes, machine minutes or in general, minutes. To produce an item, either the man or the machine or both would be required to work for a certain time. If a man works for one hour, the unit of this work is called a man hour. In other words, if 5 persons are required to work together for 20 hours to build a certain wall, the work content in building this wall is 100 man hours.

Similarly, if a machine works for 30 minutes to produce a single component, the work content in producing this component by the machine is 0.5 machine hours or 30 machine minutes.

Example: A workshop produces a certain component from rawcasting. The operations involved are fettling, machining, grinding, pretreating and painting, the operation times in each case being 8.0, 15.0, 5.0, 4.0, 5.0 minutes respectively per unit. Determine the total work content in the production of these components:

Answer: The total work content is the sum of the machine minutes of all the operations viz., $8 + 15 + 5 + 4 + 5 = 37.00$ man minutes.

4.8 Basic work content

While the above paragraph illustrates the work content of a job in general terms, its main component is the basic work content, which is the least quantum of work to be done to produce a product. Basic Work Content can be understood by the following explanatory definition.

Basic Work Content is the quantum of work to be performed for the performance of an operation at the least and irreducible time, at a standard rate of working, in which a task can be completed if everything is carried out perfectly.

Thus, the basic work content relates to the ideal situation and corresponds to the time taken to manufacture the same if the design and specifications are perfect, if the process or methods of manufacture is optimal and perfectly carried out, and if there are no losses of working time due to any causes whatsoever during the period of operation. In other words the basic work content is the absolute minimum work theoretically required to produce one unit of the product. As we can see from the chapter on work measurement later, this irreducible time gets added up by the relaxation, etc. allowances to provide for the fatigue induced onto the operative.

4.9 Excess work content or added work content

The basic work content gets added due to several imperfections in the design or the manufacturing process or other factors that may be within the control of the management or the operatives. As we are aware, there is always a difference between the ideal and the actual situations. Perfect situations never occur in practice, though they can be approached. Over and above the basic, the work content is always added up due to a variety of reasons which can be classified as:

1. Excess work content added due to imperfect designs and specifications,
2. Excess work content added due to inefficient methods and processes of manufacture.
3. Ineffective time added due to factors within the control of the management.
4. Ineffective time added due to factors within the control of the workforce.

This additional work content adds to the minimum time taken over and above that required for the basic work content.

Total working time = Basic work content + added ineffective time or

$$T_t = B_t + A_t$$

Fig. 4.3. represents diagrammatically all these four factors. While Fig. 4.4 illustrates how the excessive work content is added to the basic work content due to bad product and process design, Fig. 4.9 illustrates how the ineffective time is added to the basic work content due to factors within the control of the management and the operatives.

4.10 Excess work content added due to imperfect design and specifications

It is a very common situation in (an added) an industry where the design of a product is done without proper and complete knowledge of the functional aspects of all the design features and sometimes where the material availability situation has changed over the passage of time. These factors result in imperfect design and specifications. In accordance to the definition given in the earlier paragraph, this situation contributes to the excess work content added due to imperfect design and specifications. The under mentioned illustrations detail how the excess work content gets added up due to this factor.

- The product or its component may be so designed that it is impossible to use the more economical processes or methods of manufacture. This applies especially to the metal working industries and more particularly where large scale production is undertaken. For example, a sheet metal part may be so designed to be cut, riveted or welded instead of being pressed in one piece.

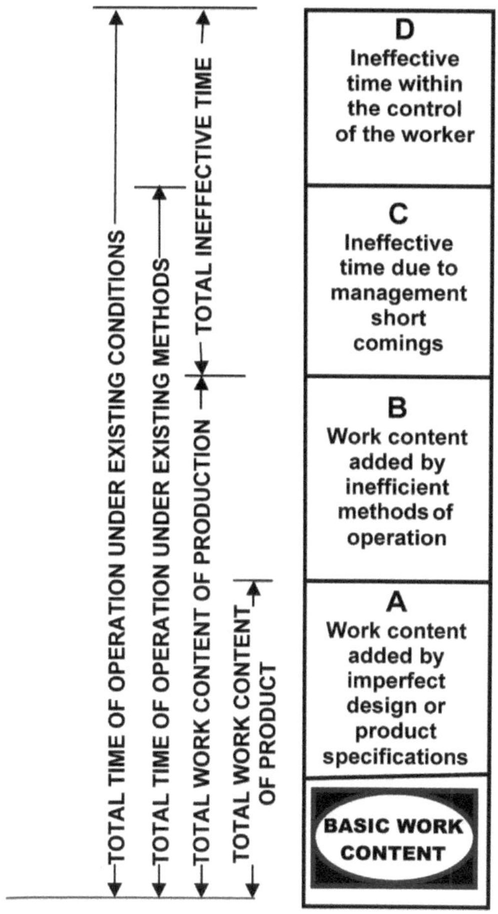

FIGURE 4.3 Factors that add to the basic work content of producing the production the product.

- Excessive product variety or lack of standardization of components may result in working in small (smaller) batches of production and may have to be done on slower and less productive machines.
- Some components of a product may be so designed to have an (an added) excessive amount of material to be removed to bring them to their final stage, may be at the time of designing, the product it is the nearest raw material size available. The examples shown in the Figs. 4.5 and 4.6 represent a certain component designed to be produced by turning out of a 50 mm diameter bar just to maintain a collar of 10 mm. As in Fig. 4.5, excess work has to be done for the removal of excess material to the tune of 60%.

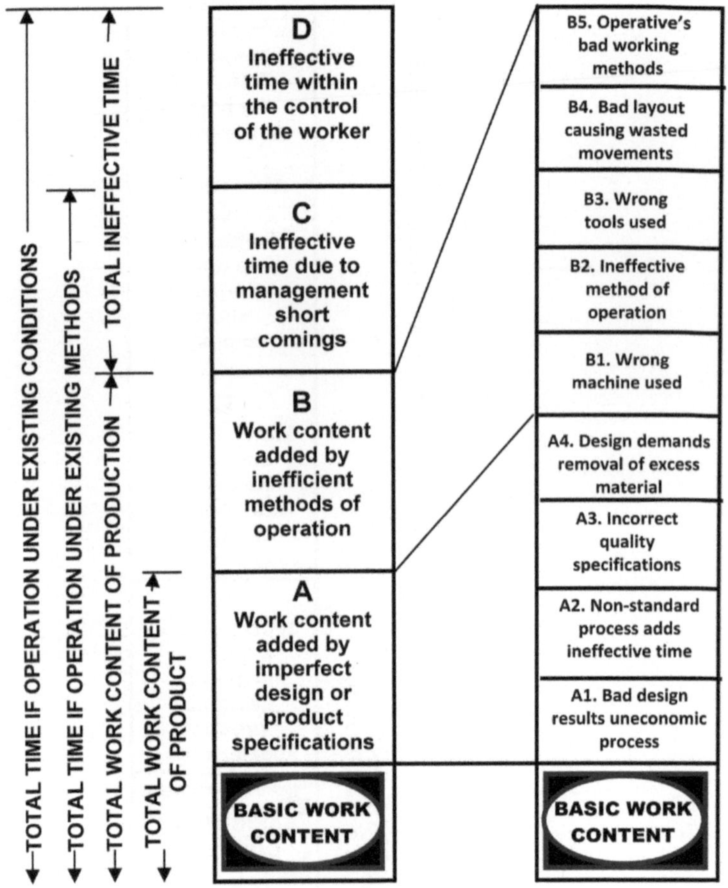

FIGURE 4.4 A: excess work content added due to imperfect design and specifications. B: excess work content added due to inefficient methods operation.

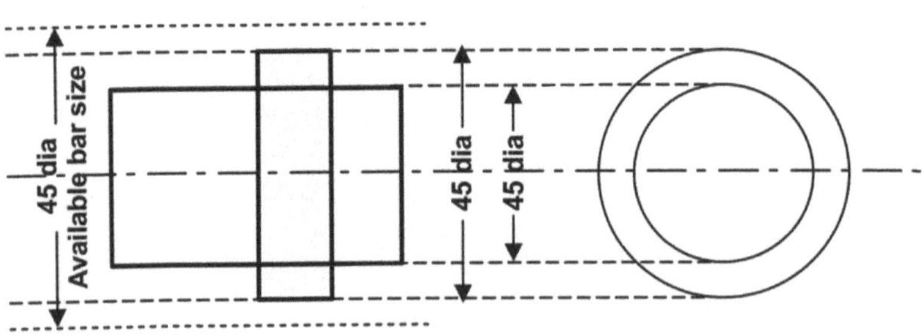

FIGURE 4.5 An example of imperfect design feature.

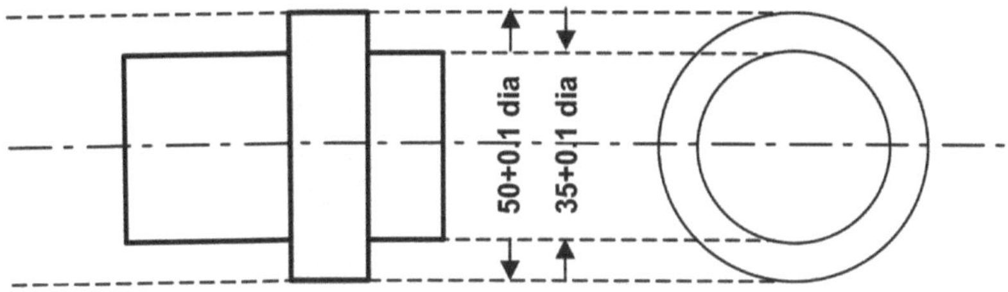

FIGURE 4.6 An example of unnecessary tolerance specification.

- Many times, higher quality standards that are not functionally justified could have been specified in the product design. For example, by specifying close tolerances to certain dimensions that do not mate with other components as in Fig. 4.6, we require additional precision machining, besides the use of costly machinery and control system.

The first step towards raising productivity and lowering the cost of production is therefore to eliminate, as far as possible, all the design and specification features that have no functional significance, but cause excess work content. This is also called Value engineering as explained further in Chapter 18 on Value Analysis.

4.11 Excess work content added due to inefficient methods and processes of manufacture

It is also a common situation in industry, where inefficient methods are specified in the process itself. This wrong selection of the process can result in excess content added due to inefficient methods of manufacture or process. The under mentioned illustrations detail how the excess work content gets added up due to this factor.

- In a machine shop a small component is being specified to be turned on a large center lathe, even though there exists a small capstan lathe on which the machining could have been done faster.
- In a textile mill 25 mm wide cotton tape is woven on a regular basis, on a loom meant for weaving a 1 m wide cloth.
- A certain bank has 136 column printers connected to its computer but all the monthly account statement is designed for 80 columns printing.
- In a machine shop a certain new component has first been machined on the shaping machine during the trial production. When it is subsequently taken up for mass production, the process is continued to be specified to be done on the shaper, even though a milling machine is available in the machine shop.
- A certain product has a drilled hole of 9 mm diameter. Though a drilling machine A of 10 mm capacity is available, a larger drilling machine B of

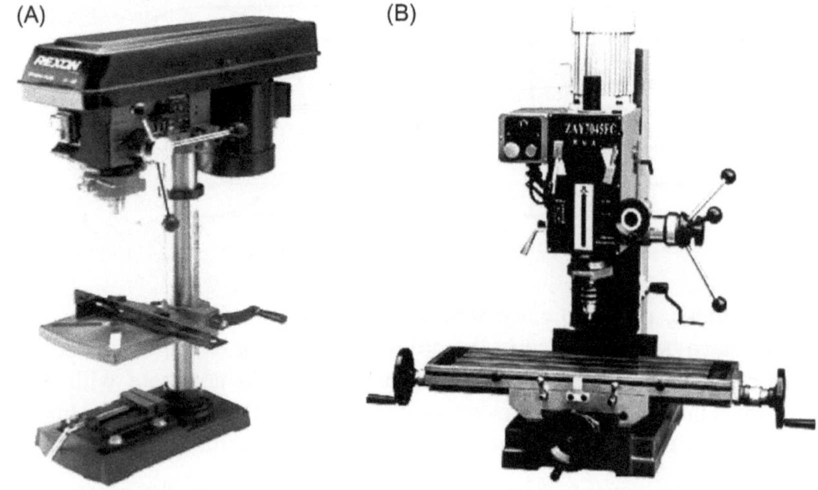

FIGURE 4.7 An example of selection of wrong machine.

FIGURE 4.8 An example of using wrong tools for blanking and punching of holes.

25 mm capacity is specified. This too, is an example of selecting a wrong machine as illustrated in Fig. 4.7.

- A certain pressed component as indicated in Fig. 4.8A is obtained by first producing the blanks in a fairly large batch size of 5000 pieces, on a 45-ton press and then punching out the three holes on a separate 10 ton press. This is an example of wrong method and tool selection, since the same could have been replaced by a progressive die to get the component in a single operation. The former is thus an example of excess work content being added due to wrong selection of the tool. Fig. 4.8B illustrates the lighter side of the use of wrong tools.
- A process which does not operate properly at correct speeds, temperatures, etc., consumes higher ineffective time, resulting in the need for excess work, over and above the basic work content which corresponds to the optimum conditions of speeds, temperatures etc.

- The layout of the plant that is not suited to the smooth flow of the material from and to the different work stations, also results in excessive work content due to additional materials handling.
- Several times the welding fixtures are not used properly or not used at all, even though they are available at site. This causes wasted efforts in positioning the components before welding as well as rejections.

The factors A and B illustrated in Fig. 4.4. together form the excess work content inherent in the production of the component under the existing specifications. This together with the basic work content forms the total work content of the product and corresponds to the minimum operation (operational) time of the product.

4.12 Excess work content added due to factors within the control of the management

There are also other factors that add ineffective time to the basic time of operation. These factors are not necessarily due to the design of the product or the process, but due to various other factors that are fully within the control of management.

This is illustrated as C in Fig. 4.9

- The excessive product variety that are scheduled for simultaneous production as a result of wrong marketing policy, adds idle time due to short runs.
- Frequent design changes cause frequent stoppages of work and rework which again increases idle work. This situation may also be due to wrong marketing policy and causes excess work content.
- In a competitive market, the car manufacturers may produce varying colors and accessories per customer's choice in a single day. But here the main body manufacturing or the engine manufacturing line is not affected. Only the painting and accessory assembly line would be varied, enabling minimal design or process changes.
- If the production planning and control systems are not properly planned and followed, the idle times of the men and machines will increase the excess work content.
- Improper inventory control results in frequent stock outs causing machine stoppages due to the material not being available at the right machine at the right time in the right quantities.
- Improper maintenance policy causes frequent plant breakdowns and machine stoppages.
- If the plant and machinery are allowed to continue to work in bad condition, due to the absence of an effective preventive maintenance policy, rework and large quantities of scrap results, besides high idle times.

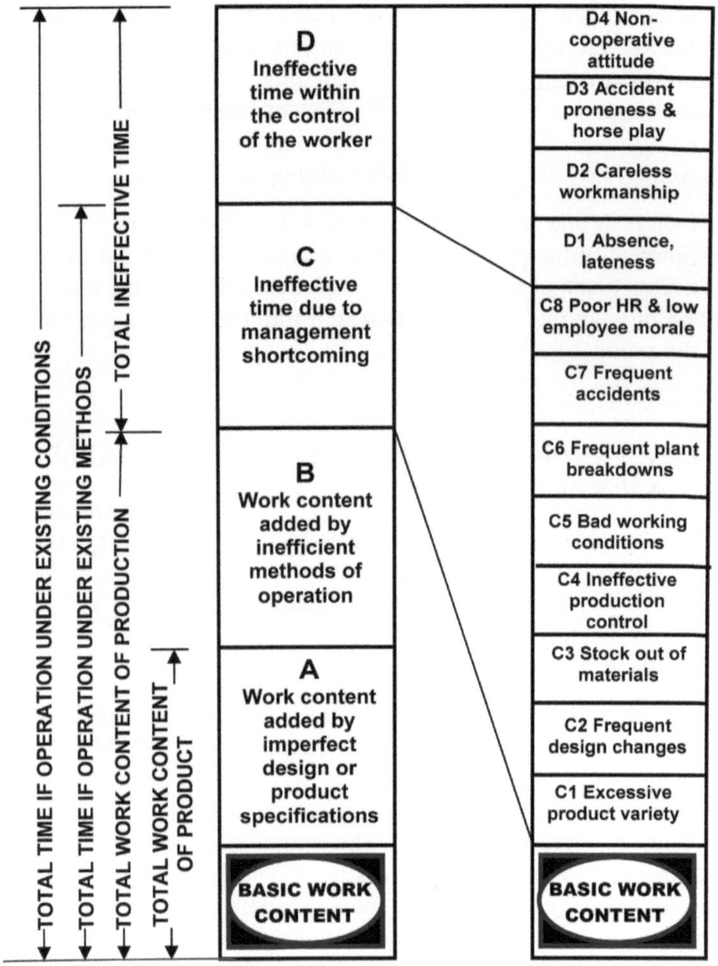

FIGURE 4.9 C: excess work content added due to management shortcoming. D: excess work content added due to factors within the control of the worker.

- Bad working conditions like poor lighting, lack of ventilation, etc., cause reduced production output and in increased scrap production, adding to the enforced rest time for the workers in addition to what is normally taken.
- Poor attention given by the management towards proper safety training and maintaining ideal safety conditions causes lost time due to accidents, both resulting in plant stoppages and absenteeism.

4.13 Excess work content added due to factors within the control of the work-force

While the various illustrations, detailed above indicate the factors beyond the control of the workforce, there are some other factors which are purely under their control, that cause excess work content in terms of ineffective time added. This is illustrated as D in Fig. 4.9.

- Absenteeism and late-coming without proper cause and also deliberately not starting the work immediately after starting the machine, are some of the more common factors that increase the work content.
- Even after coming in time, some workers deliberately slow down their normal pace of work. This may be due to various reasons including tool down or go slow strikes. Such practices only add up to the excess work content.
- If a worker causes excessive scrap or rejections by careless workmanship, it results in wastage of material and time and sometimes in rework.
- Even after the provision of guards, goggles, etc., the workers sometimes fail to observe the safety regulations like not using the goggles while grinding or operating the press machine without a guard or indulging in horseplay. Accidents are caused in these cases resulting in wasted time besides other losses.

These factors are under the control of the workforce. This can be avoided by better training and human relations policies and by uplifting the employee morale

In general far more ineffective time is added due to the factors within the control of the management than due to the workforce. This is especially true in case of large and medium industries manufacturing complex products.

4.14 Summary of the factors that add to the ineffective work content to the production

We can summarize as under the various factors that add to the ineffective work content to the production.

A. Work content added by ineffective design or product specifications

A1. Bad design results uneconomic process
A2. Non-standard process results ineffective process
A3. Incorrect quality specifications
A4. Design demands removal of excess material

B. Work content added by inefficient methods of operation

B1. Wrong machine used
B2. Ineffective method of operation
B3. Wrong tools used
B4. Bad layout causing wasted movements
B5. Operative's bad working methods

C. Ineffective time due to management shortcomings

C1. Excessive product variety
C2. Frequent design changes
C3. Stock out of materials

C4. Ineffective production
C5. Bad working conditions
C6. Frequent plant breakdowns
C7. Frequent accidents
C8. Poor HR & low employee morale

D. Ineffective time within the control of the worker

D1. Absence, lateness
D2. Careless workmanship
D3. Accident proneness & horseplay
D4. Non-cooperative attitude

These factors can be tabulated as below in Table 4.2.

TABLE 4.2 Factors adding excess work content and the remedy.

Ref No	Factors adding excess work content	Industrial engineering techniques
A. Work content added by imperfect design or product specifications		
A1	Bad design results uneconomic process	Effective Product development with Value Analysis
A2	Non-standard process adds ineffective time	Materials utilization Analysis to reduce and utilize waste.
A3	Incorrect quality specifications	Total Quality Management
A4	Design demands removal of excess material	Effective Material layout planning
B. Work content added by inefficient methods of operation		
B1	Wrong machine used	Better process planning and layout
B2	Ineffective method of operation	Method study
B3	Wrong tools used	Process planning
B4	Bad layout causing wasted movements	Plant Layout and Materials handling
B5	Operative's bad working methods	Micro motion study
C. Ineffective time due to management shortcomings		
C1	Excessive product variety	Standardization and codification
C2	Frequent design changes	Group technology
C3	Stock out of materials	Inventory control

(Continued)

TABLE 4.2 (Continued)

Ref No	Factors adding excess work content	Industrial engineering techniques
C4	Ineffective production control	Production Planning & Control
C5	Bad working conditions	Industrial safety incorporating better working conditions improves morale
C6	Frequent plant breakdowns	Condition based preventive maintenance
C7	Frequent accidents	Job Hazard Analysis
C8	Poor HR & low employee morale	Proper HR management and personnel policy
D. Ineffective time within the control of the worker		
D1	Absence, lateness	Job interest creation
D2	Careless workmanship	The job training
D3	Accident proneness & horseplay	Effective training and awarding accident free records
D4	Non-cooperative attitude	Total Employee involvement

4.15 Productivity improvement by group technology

Group technology is a manufacturing technique in which parts having similarities in geometry, manufacturing process and/or functions are manufactured in one location using a small number of machines or processes. This helps us in producing smaller quantities of several components in a single setting or with minor variation in the turning tool position. This will give much higher productivity than producing each component with different settings.

4.16 How industrial engineering techniques help in minimizing the excess work content and the ineffective times

Having understood the several factors that add to the basic time, we can attempt to minimize them one by one as listed in Table 4.2. In addition, Fig. 4.10 summarizes these techniques and signifies that these techniques, especially method study wing of work study, in increasing the productivity of every manufacturing organization, thereby contributing to the prosperity of the nation.

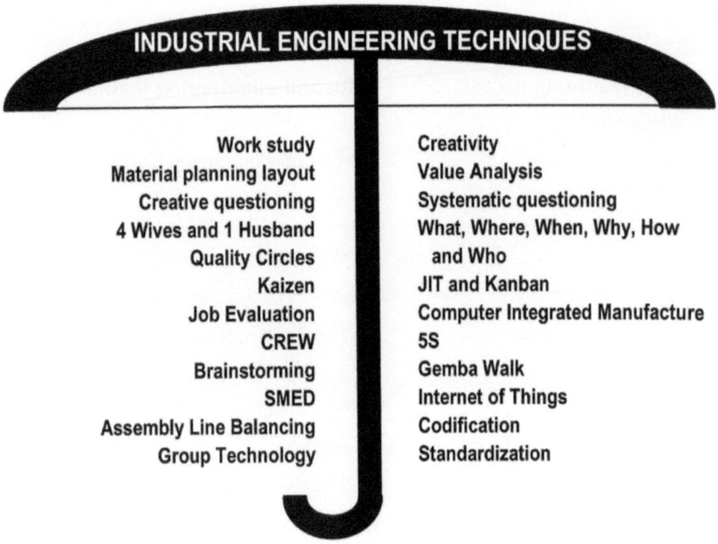

FIGURE 4.10 The umbrella of industrial engineering.

It may however be noted that none of these techniques can be applied in isolations since most of them are interrelated. For example

- A well-conceived codification pregame would result in variety reduction, in addition to the standardization program.
- A Standardization results in variety reduction of the manufactured and/or the purchased components, aiding in material control and production control.
- Product research and development keeps in mind value analysis, standardization, maintenance planning, etc. at the time of developing the product and process.

4.17 Impact of iot and AI on productivity enhancement

Today's buzzword is 'Industry 4.0', metaphoric representation of the fourth industrial revolution brought out by the IoT (Internet of Things) and cyber application by way of Artificial Intelligence in industrial operations. To give an impetus to the German Manufacturing, the German Government initiated Industry 4.0 as a project for the promotion of the computerization at advanced levels in manufacturing during the first decade of 21st century. The original industrial revolution of 1760 was considered as industry 1.0 or the first industrial revolution. The extensive use of electrical power in place of steam power for mass production during the early and mid-20th century was considered as industry 2.0, or second industrial revolution while the extensive use of IT and electronics for automation in mass production during the latter half of 20th century was considered as industry 3.0 or third industrial revolution.

4.18 Conclusion

In this chapter, we have summarized the different factors that contribute to the excess work content during the production of goods, resulting in loss of productivity. Against each such factor, some of the industrial engineering techniques that can be applied to prevent such addition of excess work content are cited. However, it may be noted that each of these techniques cannot be applied in isolation, but by treating the enterprise as a whole. Work study uses these types of techniques, whether manual or computerized, to develop new methods of work as can be seen in the later chapters.

Criteria questions

1. Cite some of the symptoms of low productivity. (4.1)
2. Illustrate the existence of hidden costs. (Fig. 4.1)
3. Distinguish between Internal and External factors affecting (effecting) the Productivity. (4.3)
4. Distinguish between hard and soft factors. (4.3)
5. Distinguish between basic Work Content and added ineffective work content. (4.9)
6. Illustrate the four ways by which the ineffective work content gets added. (4.9)
7. How does Group Technology Improve Productivity? (4.16)

Further reading

1. ILO. *Introduction to work study*. 3rd ed. ILO; 1979.
2. Propekenko J. *Productivity management*. ILO; 1987.
3. Faraday JE. *The management of productivity*. Bureau of Indian Standards; 1971.
4. Kiran DR. *Production planning and control, a comprehensive approach*. BS Publishers; 2017.
5. Kiran DR. *Productivity & work study in industry*. Work Study Seminar, University of Dar Es Salaam; 1980.
6. Kiran DR. *Programmed instruction manuals for NCES*. NPC; 1985.
7. Carson G, editor. *Production handbook*. Arnold Press; 1978.

Chapter 5

System approach to productivity

5.1 Development of system approach

In the past the departmental managers tried to solve their problems by considering them as isolated situations, independent of other activities of the organization. For example, if a certain product manufacturer noticed a sales decline and traced it to the lack of aggressive sales effort, it was treated purely as a sales management problem and action was taken against that particular salesman or the department. The contribution of other departments to this effect like the quality control, design, credit policy, or the advertising, were never given much thought. Life study of any object must rely on the method of analysis involving the simultaneous variations of mutually dependent variables. As early as the fifties, a new integrated approach which considers the management in its totality has been developed and this is called the system approach.

5.2 What is a system?

We are surrounded by and live in several systems. In fact, man is a system by himself. Some systems are natural like the planetary systems, animal systems, and environmental systems. Others are man-made, like the business systems, production systems, material handling systems, social systems and total quality management systems. Whichever is the system it is basically characterized by three components that form the input-process-output system per Fig. 5.1, which comprise of

- A set of two more elements forming the input
- Seeking a common goal operating on certain data or matter or energy, forming the process, and
- Yield matter or data in a time frame, forming the output

5.3 Definition of a system

Originating from the Greek word sustema, the word system has three definitions in Oxford dictionary as below,

Work Organization and Methods Engineering for Productivity.
DOI: https://doi.org/10.1016/B978-0-12-819956-5.00005-4

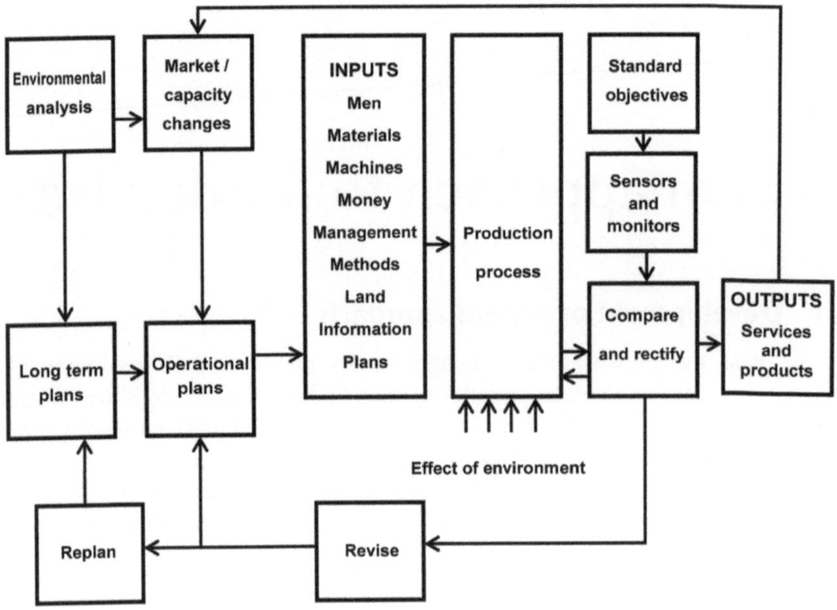

FIGURE 5.1 The complete production system.

1. Group of things or parts working together, like the digestive system, railway system or the control system or the computer field, a group of related hardware units or programs or both. Its adjective is systemic, meaning related to a system as a whole.
2. A set of ideas, theories or principles or the social order like a political system or a government system.
3. Orderliness or methodical or an organized manner of working. Its adjective is systematic and the word systematize means arranging things according to a system.

American Management Association in its AMA Management Handbook explains these three groups of definitions as follows

1. The first group related systems to living creatures, including the human beings, emphasizing on the impact of personality and behavioral sciences.
2. The second group emphasizes on the general meaning of the system as a set of two or more elements like people, things or concepts, which are related or joined together to achieve a common goal, operating on data, matter and energy, to yield a result with a time reference. For example, a system of Government, a system of measurement or a system of classification.

3. The third group of definitions connects systems with specific science techniques like philosophy, mathematics or biography.

Whatever may be the definition, a system approach is a purposeful and powerful means for accomplishing an objective. It provides the management with an analytical framework with which it can identify, describe and interrelate the process and the components that make up a particular system as explained and illustrated in the following paragraphs. The study of any object or process must rely on the methods of analysis involving the simultaneous variations of mutually dependent variables.

5.4 Components of a system

Systems are most often represented in flow charts and block diagrams. The major components of a system are input, process, and output. Fig. 5.1 illustrates the process in its simplest form.

5.4.1 Input

Input to a system can be classified basically into 3 categories. The first comprises of the physical inputs, including the materials that flow through the conversion process where work is performed on them. The second is the non-physical materials, but required for the performance of the process, like the management, money, capital, energy, labor and land. The third category is the environment that affects the system's operation. The determination of the amount, placement, timing, and types of these inputs will have an impact on the conversion facility, whether it is a factory or a hospital or an office.

5.4.2 Conversion process

The second basic part of the system is the conversion process through which the inputs flow to produce the desired outputs. To be effective and efficient, systems must be designed so that the correct process acts on the inputs at the proper time.

5.4.3 Output

The third major component of the system after the conversion process is the output, which comprises of the desired accomplishment of the system. In an automobile industry, the output is the number of completed cars of a desired quality produced within a specified time frame. This input → process → output elements would exist whatever may be the situation, whether agriculture or educational, etc., as explained in the following paragraphs.

5.5 Types of systems

Some specific cases of systems can be explained below, illustrated by the following figures.

1. **Business organization system:** A group of people with a vision and mission, human resources, stakeholders, consultants, etc. (input), by performing production, design, finance, marketing, etc. functions (process) towards achieving the goal of optimal profit for the business (output). (Fig. 5.2).
2. **Manufacturing system:** A group of men working on machinery and materials with the aid of money and a good management, (5 M's of input), working as per work-methods, schedules and specifications (process) to yield the specified products and information by the date the customer wants them (output) (Fig. 5.3)
3. **Management information system:** A group of people with a set of manuals and data processing and recording equipment (input), working for selecting, storing, processing and retrieving data (process) to yield the information needed by the managers (output) (Fig. 5.4).
4. **Agricultural system:** In an agricultural situation, the men by use of implements, fertilizers and land (inputs), till, plant and grow (process) to achieve the agricultural produce (output) as in Fig. 5.5.

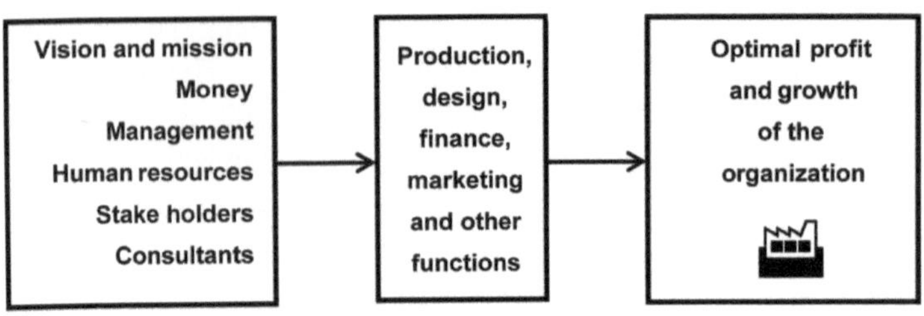

FIGURE 5.2 Business organization system.

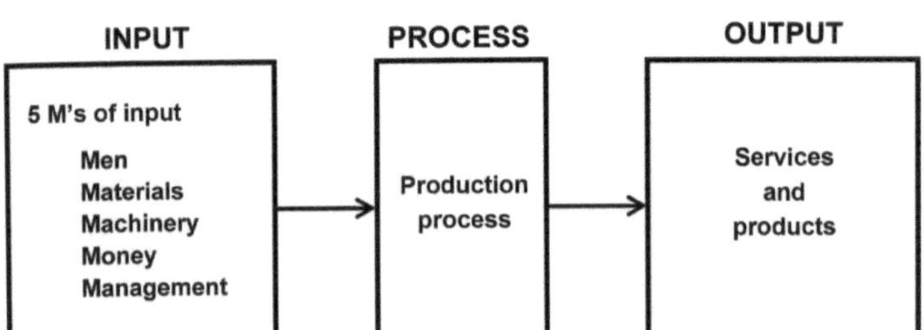

FIGURE 5.3 Manufacturing system.

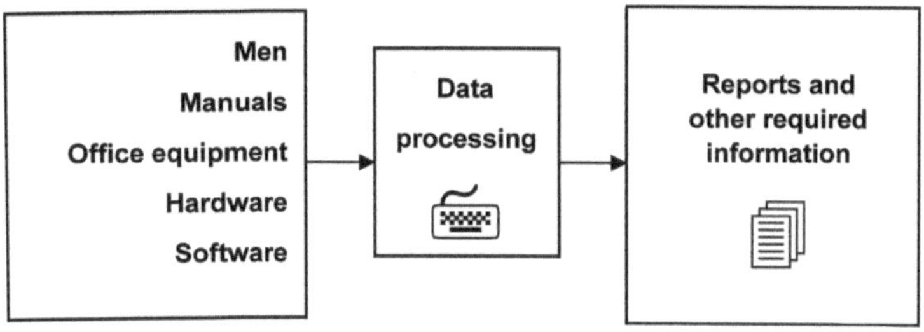

FIGURE 5.4 Management information system.

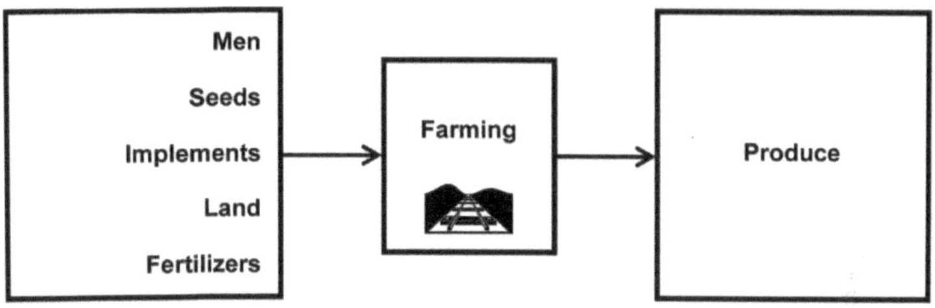

FIGURE 5.5 Input - output process in agricultural situation.

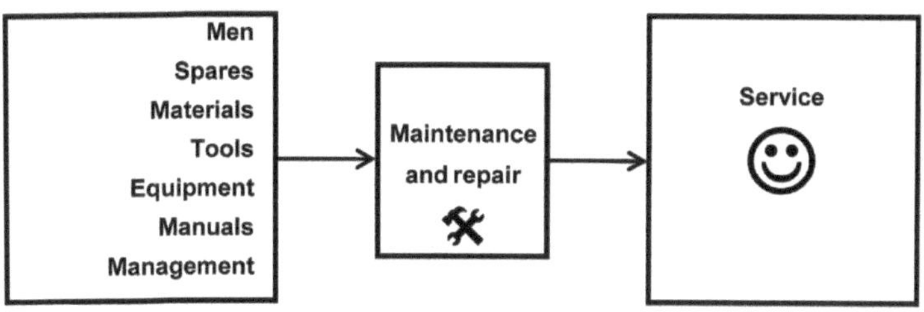

FIGURE 5.6 Input - process- output in automobile repair situation.

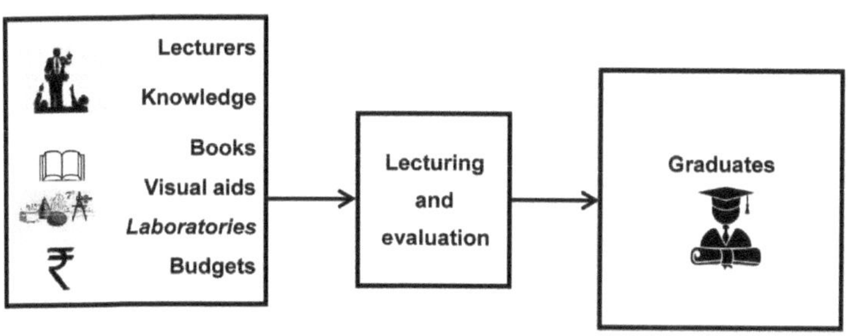

FIGURE 5.7 Input - output process in educational institution.

In an automobile repair shop, the inputs are the men, spares, tools etc,. (inputs) are utilized to do repair work on vehicles (process) to achieve vehicle reconditioning and other services (output) as in Fig. 5.6.

In an educational institution, the inputs are the lecturers and the expertise, knowledge of the teachers, the textbooks, the visual aids, etc. (inputs) perform lecturing, training and evaluation of the students (process) to achieve the enlightenment of the students (output) as in Fig. 5.7.

5.6 Elements of control in system approach

Just as goals or objectives are needed in the design of a system, control is required in its operation. The purpose of the control is to maintain the quality and quantity of the output so that it meets the goals of the system. In the ideal case the control would be self-regulating and holds the output at desired levels without outside intervention. The control has the following elements.

1. A well-defined objective
2. An output standard
3. A measuring device
4. A feedback signal and
5. A corrective course of action

Unfortunately, in most management situations not all of the elements of control are present or not precisely defined. For example, consider a production process in which the standard of the units of output per hour has been established for a product. However, when the output from the process is monitored, it is found that 80% of the units only are produced per hour. In this situation, there is an objective, an output standard, a measuring device and a feedback signal, but no corrective action has been specified.

5.7 Environment

A system's environment is composed of all activities outside the system which, if changed, will affect the system or which will be changed by any changes in the system itself. In this context, the general level of economic activity can be considered as environmental for most production activities, since any change in it will affect the production levels.

5.8 Open and closed systems

Every system has a boundary which functions to maintain proper relationships between the system and the environment. A closed system has no environment around the system. No outside systems have any effect on this system and the inputs and outputs have definite mathematical relationships like a chemical process inside a hermetically sealed container. (Fig. 5.8A).

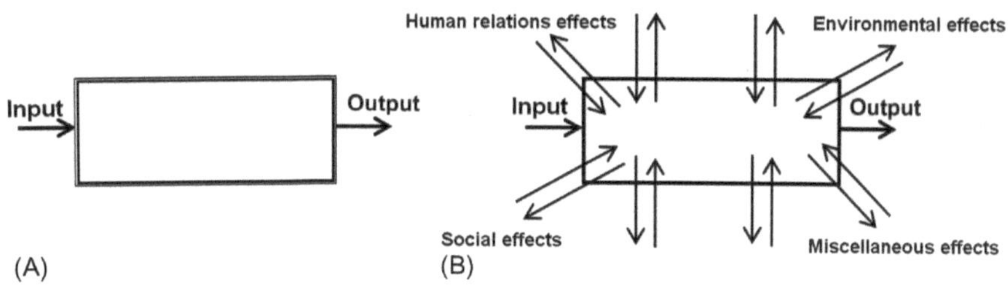

FIGURE 5.8 (A) Closed and (B) open systems.

The open system has an environment with which it relates exchanges and communicates, like a human system or a production system. (Fig. 5.8B).

5.9 Systems and subsystems

Because of the complexity of most real systems, it is extremely difficult to work with or to understand the entire system. In order to overcome this problem, systems are broken into smaller subsystems and still smaller sub-subsystems to comprehend them better. For example, the whole universe is a system, while the solar system is its subsystem. We may still break it down to the earth system, world system, a group of nations, a country system, a social system, man as a system, blood circulation system, etc., as illustrated in Fig. 5.9 shown alongside.

5.10 Relationship between the systems and subsystems

Each system and its subsystems are mutually related, some more and some less, some directly and some indirectly. This relationship is in the context of the whole and is complex. Any change in one part would affect the other to a varying but a predictable degree. Fig. 5.10 illustrates the relationship between the systems while Fig. 5.11, the different generations of subsystems.

5.11 Combination of subsystems

Two or more subsystems can be integrated in either series, or in parallel or in combination as illustrated in Fig. 5.12.

5.12 The management cube

The management activity is influenced by three aspects

1. The management processes, including the planning, controlling and execution,
2. The management functions like design, planning, maintenance, marketing, purchasing, personnel and finance,

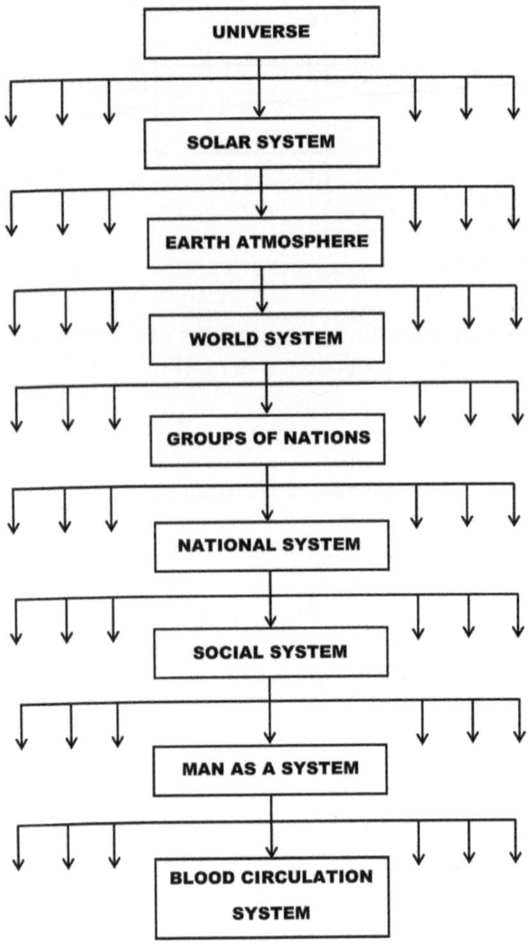

FIGURE 5.9 Division of systems and subsystems.

3. The management level, like operatives, supervisors, engineers, managers, directors, president and Chairman.

These three activities can be illustrated by Fig. 5.13, which is called the Management Cube. It shows how the managers of different functions perform the processes at different levels.

5.13 Planning pyramid

The management cube considers the same number of managers at each level. But in fact the number of persons in each level increases as we go down the hierarchy levels, because of the concept of span of control. For example, the planning of the vision and mission or the policy making is done at the

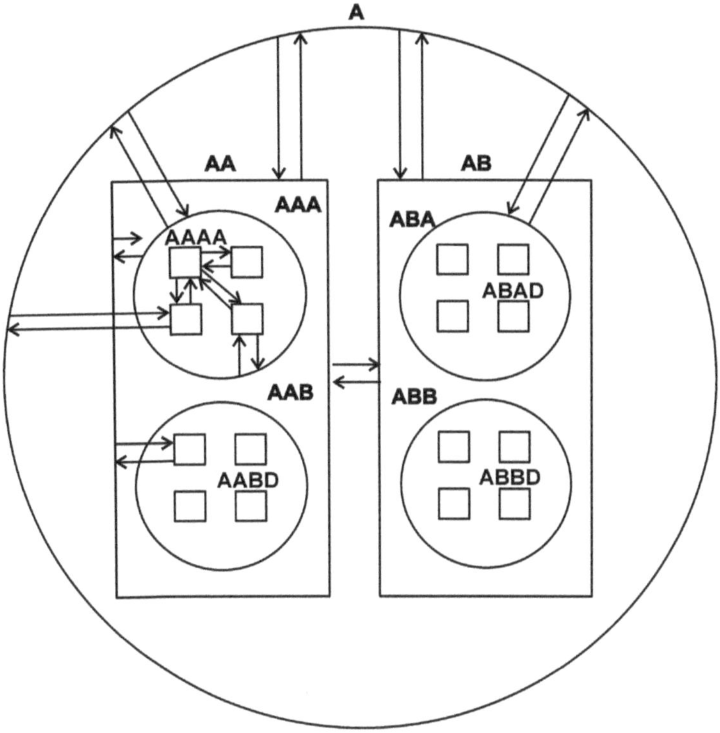

FIGURE 5.10 Interrelationships between systems and subsystems.

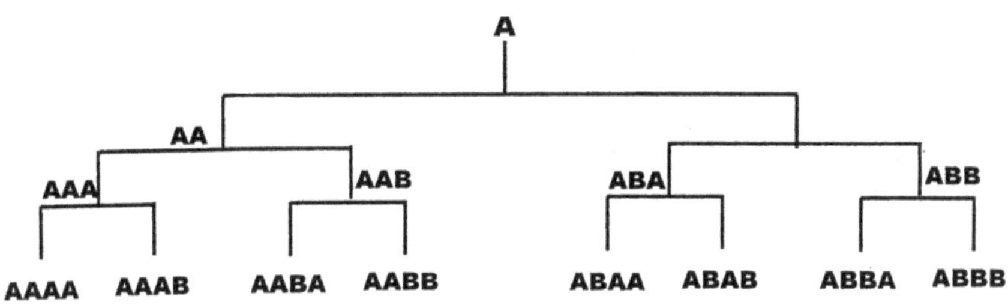

FIGURE 5.11 Generations of subsystems.

Chairman level, while the day-to-day operational planning is done at the Manager level. As one ascends higher in the hierarchy, the level of planning is escalated and simultaneously the number of individuals responsible narrows down.

These activities are better represented by a pyramid signifying this aspect and showing what type of planning is done at each level in the organization, as illustrated in Fig. 5.14.

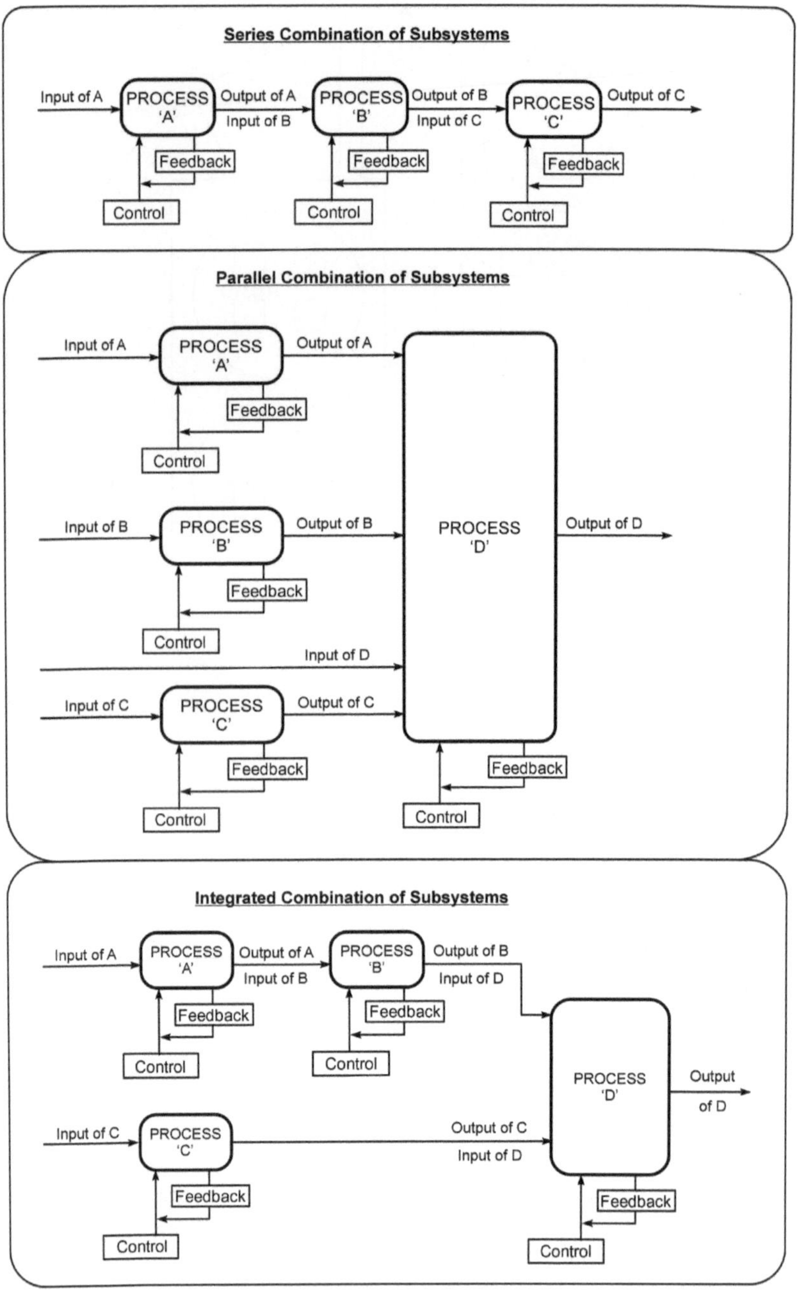

FIGURE 5.12 Combination of subsystems.

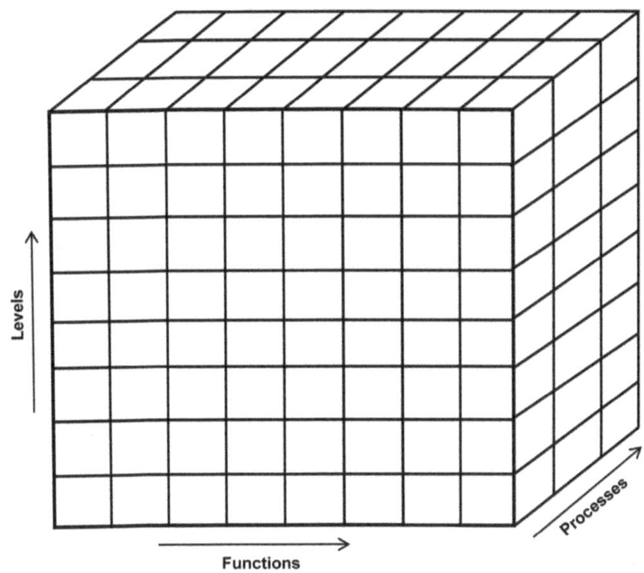

FIGURE 5.13 Management cube.

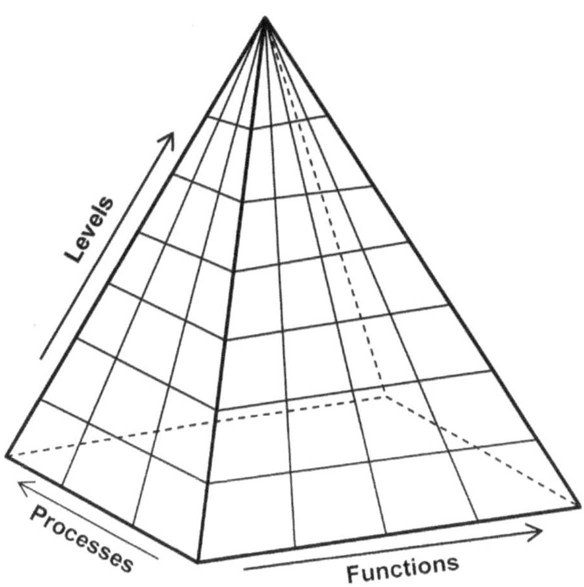

FIGURE 5.14 Planning pyramid.

5.14 Decision theory

The day to day administration of the affairs of any business enterprise requires an endless sequence of decisions. Decision making is an integral part of a team, given the responsibility of meeting and discussing among themselves towards the achievement of a goal. They may either be routine or unusual, demanding the manager's closest attention and intelligence.

In fact the terms, to decide, to determine, to settle and to conclude, are generally used as synonyms but we may distinguish between them as follows:

- To decide implies the bringing to an end any doubt, dispute, vacillation or wavering between choices, by making up one's mind on what to do, the course of action, etc.
- To determine, in addition to the above, suggests or fixes precisely the form, character, functions, scope, etc.
- To settle stresses finality in decision, often the one arrived at by arbitration process and implies the termination of all doubts and controversy.
- To conclude means to decide after careful investigation and reasoning.

In the past when it was mostly a seller's market, profit maximization was perhaps the sole motto of an enterprise, and accordingly all decision makings were simple. But today the manager is influenced by a hoard of factors.

1. Increased regulation of business activity by the local or National governments.
2. Growing separation of ownership and management in industry, that is the management becoming more and more professional.
3. Steadily increasing importance given to high level education for managers.
4. Development of new processes and equipment.
5. Expansion of the market areas in view of improved communication and transport logistics.
6. Increased competition from other suppliers.
7. Growth of labor unions and more stringent labor laws.
8. Growing social pressure.

The following paragraphs detail the decision theory with regards to identifying its rationality, uncertainties and other issues relevant in a given decision, as well as the resulting optimal decision.

5.15 Problem analysis vs. decision making

Wikipedia distinguishes Problem analysis from decision making as follows. Traditionally problem analysis must be done first, so that the information gathered in that process may be used towards decision making.

Problem analysis

- Analyze performance, what should the results be against what they actually are.
- Problems are merely deviations from performance standards.

- Problem must be precisely identified and described.
- Problems are caused by a change from a distinctive feature.
- Distinguish between what has and hasn't been affected by a cause.
- Analyze the relevant changes to identify the Causes to problems.
- Once you identify the cause, you can deduce the problem with reference to the solution.

Decision making

- Objectives must first be established.
- Objectives must be classified and placed in order of importance.
- Alternative actions must be developed.
- Evaluate the alternatives against all the objectives.
- The alternative that is able to achieve all the objectives is the tentative decision.
- Evaluate the tentative decision against your objectives.
- The decisive actions are taken, and care is taken to prevent any adverse consequences from becoming fresh problems themselves, in which case problem analysis and decision making shall start all over again.
- Determine an optimal production plan, by adapting steps that are generally followed which result in a decision model.
- In a situation featuring conflict, role-playing is helpful for predicting decisions to be made by involved parties.

5.16 Characteristics of decision making

While paragraphs 5.12 to 5.15 illustrate the systems approach to decision making, the following paragraphs of this chapter illustrate the aspect of decision making as relevant to the manager.

Decision making involves the following characteristics

- Decision making is a process of selection and the aim is to select the best alternative.
- Decision is aimed at achieving the objective of an organization if it is made in the organizational context.
- It involves evaluation of available alternatives since only by this evaluation can one know the best alternative.
- Decision making is a mental process and the final decision is made after thoughtful consideration.
- Decision making involves a certain commitment. This commitment may be for a short run or long run, depending upon the type of decision.

5.17 Situations under which decisions are taken

Decision making under certainty, is when we know with confidence what will occur. In this case we have to consider only one possibility of occurrence for every alternative and decision making is very simple.

Decision making under uncertainty, is when there are a number of alternatives and no past data is available to compute or estimate the probability of occurrence. The estimation of market demand for new and untested products is an illustration.

Decision making under risk, is when there are a number of possible alternatives and the probability of occurrence of each alternative, based on past data and past experience is known. This is controlled by the theory of probability and the result may either turn out to be correct or wrong depending upon the accuracy of the data or the evaluation procedure.

Decision making under conflicts, is when the different individuals involved directly or indirectly in decision making have contradictory opinions or data. The discussion among the team members for arriving at a decision is an illustration.

An Illustration on the lighter side for the decision making

A sea captain on a rough sea may take a decision depending upon his attitude. An optimist expects the weather to change and takes no action. A pessimist complains about the wind, prays for the heavenly abode (moksham) for all and keeps the crew on tender hooks. A realist adjusts the sails, takes control of navigational aids and takes emergency precautions.

5.18 Classifications of decisions

There are several ways of classifying decisions in an organization.

Organizational and personal decisions: In an organization when an individual takes decision as an executive for the organization they are called organizational decisions. The authority for taking such decisions can be delegated from a superior to a subordinate. Such decisions effect directly the functioning of the organization. On the other hand, an executive can also make personal decisions that affect his and his family's personal life, though sometimes these decisions may affect the organization also. In this case the decision-making power cannot be delegated.

Routine and strategic decisions: Routine decisions are taken in the context of day to day operation of the organization. They are mostly repetitive in nature. They do not require much analysis and evaluation and can be made quickly. Authority of taking these decisions is generally delegated to the lower level of the workforce. Strategic decisions are primarily done for future periods. They affect the organizational structure, objectives, facilities, finance, etc. These decisions are mostly non-repetitive in nature and are taken after careful analysis and evaluation of various alternatives and are generally taken at a higher level of management.

Policy and operative decisions: This classification is generally similar to the above, except that they refer more to the factory level production planning decisions. Decisions like plant location, plant layout, volume of production, machinery selection, sale and purchase decisions can be called policy decisions. These are sometimes published as policy manual to become the basis for other operational decisions, which are production line decisions like the scheduling, routing etc. can be called operational decisions.

Programmed and non-programmed decisions: Programmed decisions are generally repetitive in nature and are taken within the broad policy structure. They have short run impact and are taken by lower level managers. Non-programmed decisions are those taken whenever specific problems arise unexpectedly due to certain circumstances. They are generally non-repetitive in nature.

Individual and group decisions: This classification is based on the persons involved in the decision-making process. Individual decisions are taken by a single person, generally the head of the institution, with or without consulting others who are affected. These are taken in the context of routine programmed decisions where the analysis of various alternatives is simple and for which broad policy manuals are provided. Sometimes important non-programmed decisions also are taken by individuals. Group decisions are taken by groups or teams constituted for this purpose. These decisions are in general very important to an organization. They have certain positive values such as greater participation of individuals and quality of the decision, but at the same time have certain negative values like delay in decision making and difficulty in fixing responsibility for the decision-making process and the follow up action.

5.19 Different approaches to decision making

The different approaches adapted for managerial decision making offer an insight into the theory of decision making useful in teamwork.

Intuitive decision making: In history, there have been specialist decision makers like the tribal chiefs, medicine men, priests and kings who depended mostly on their haunches or intuitions based on their training and experience. They actually involve unstructured gathering and classifying of historical data, followed by subjective evaluation.

Trial and Error decision making: This is an adaptation of the intuitive decision making, when the problem is subjectively isolated, defined and a course of action is selected and followed. If the results are favorable, the decision is allowed to stand. However, if the results tend to become failures, a second course of action is followed.

Follow the leader decision making: Here the decision is based on similar issues by other firms or an accepted trade practice, emphasizing that 'if it works fine with them, why not for us?' While this approach minimizes the risk, its main disadvantage lies in the fact that solutions depend very much upon the specific situations, which may vary from case to case and

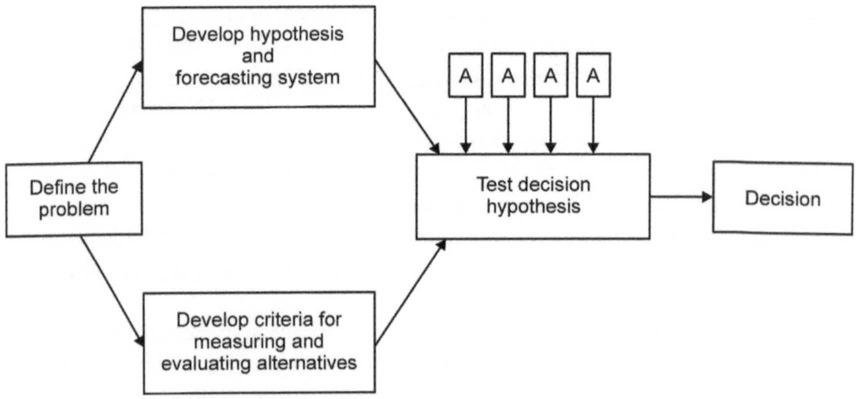

FIGURE 5.15 Systematic decision-making process.

individual to individual. Besides the variables in the same situation may change from time to time.

Scientific decision making: Here each problem is analyzed considering each of the alternative solutions. This process of analysis has been adapted in the physical services for centuries. The main 4 steps are

- Define the problem,
- Develop a hypothesis,
- Test the two hypotheses and
- Prove or disprove the hypotheses.

This is better understood by Fig. 5.15.

5.20 Systematic decision making

This is similar to the scientific decision making, excepting that a holistic and systematic approach is given considering all the related factors as part of whole system. The steps to be undertaken for this approach would be

1. Outline your goal and outcome.
2. Gather data.
3. Develop alternatives (i.e., brainstorming)
4. List pros and cons of each alternative.
5. Make the decision.
6. Immediately take action to implement it.
7. Learn from and reflect on the decision.

5.21 Information flow

As just noted to effectively control a system, data relevant to the activity must be transmitted to the decision maker. This is true whether the system is self-regulating or requires management for decision making.

For analysis, the system data flow can be divided into three classes

- operating data
- control data, and
- planning data

Operating data

Operating data include all the routine information that must be generated within the system, in order to keep it in motion. Some types of information that must be kept to operate a business are the reports, purchase orders, payroll records etc.

Control data

The second type of data is control (or feedback) information. Since systems are designed to achieve specific objectives, they must be monitored continuously to make sure that they are meeting those objectives satisfactorily. This data reflecting the status of operation must be constantly reviewed. For example, in a production of an item a specified number of labor hours, often the management needs to know if this standard is being met or if there is a significant deviation.

Planning data

The last type of data that should be made available is planning data. This includes all the information that will be required by management to solve anticipated future problems.

5.22 Bias in decision making

Wikipedia lists the following instances of bias that creeps into to the decision-making process, which should be kept in mind by the decision makers.

1. Selective search for evidence - We tend to be willing to gather facts that support certain conclusions, but disregard other facts that support different conclusions.
2. Incomplete search for evidence - We tend to accept the first alternative that looks like it might work.
3. Inertia - Unwillingness to change thought patterns that we have used in the past in the face of new circumstances.
4. Selective perception - We actively screen-out information that we do not think is salient.
5. Over-optimism- Many times we tend to want to see things in a positive light and this can distort our perception and thinking.
6. Choice-supportive bias occurs when we distort our memories of chosen and rejected options to make the chosen options seem relatively more attractive.

7. Recency - We tend to place more attention on more recent information and either ignore or forget more distant information.
8. Repetition bias - A willingness to believe what we have been told most often and by the greatest number of different of sources.
9. Anchoring and adjustment - Decisions are unduly influenced by initial information that shapes our view of subsequent information.
10. Group think - Generally group discussions leads to better brainstorming (two heads are better than one) but the unwanted peer pressure may not conform to the opinions held by the group as a whole.
11. Source credibility bias - We reject something if we have a bias against the person, organization, or group to which the person belongs. We are inclined to accept a statement by someone we like.
12. Incremental decision making and escalating commitment - We look at a decision as a small step in a process and this tends to perpetuate a series of similar decisions. This is also called as a slippery slope.
13. Attribution asymmetry - We tend to attribute our success to our abilities and talents, but we attribute our failures to bad luck and external factors. We attribute other's success to good luck, and their failures to their mistakes.
14. Role fulfillment or Self-fulfilling prophecy - We conform to the decision-making, expectations that others have of someone in our position.
15. Underestimating uncertainty and the illusion of control - We tend to underestimate future uncertainty because we tend to believe we have more control over events than we really do. We believe we have control to minimize potential problems in our decisions.

5.23 Decision tree

Decision Tree is a pictorial representation of a decision situation, normally found in discussions of decision-making under uncertainty or risk. It shows decision alternatives, states of nature, probabilities attached to the state of nature, and conditional benefits and losses.

Investopedia Financial Dictionary defines the decision tree as a management tool to clarify and find an answer to a complex problem. The structure allows users to take a problem with multiple possible solutions and display it in a simple, easy-to-understand format that shows the relationship between different events or decisions. The furthest branches on the tree represent possible end results. The tree approach is most useful in a sequential decision situation and Fig. 5.17 illustrates the logical testing procedure.

If a process consisting of a group of units connected in series as per Fig. 5.16 fails, then individual units have to be checked for failure since the exact unit which has failed is not known, the decision tree illustrated in Fig. 5.17 would help.

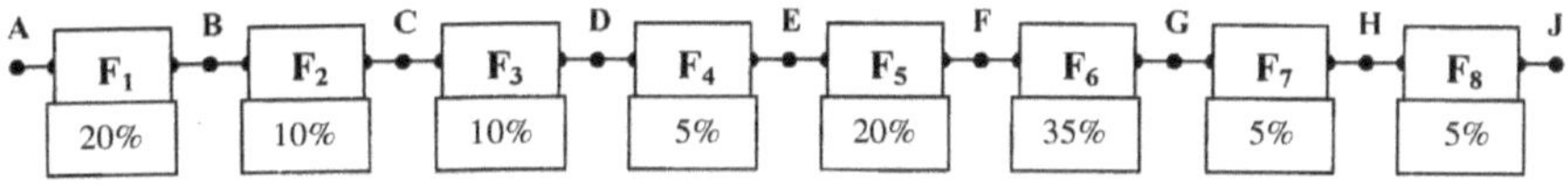

FIGURE 5.16 A simple series circuit with known failure probabilities.

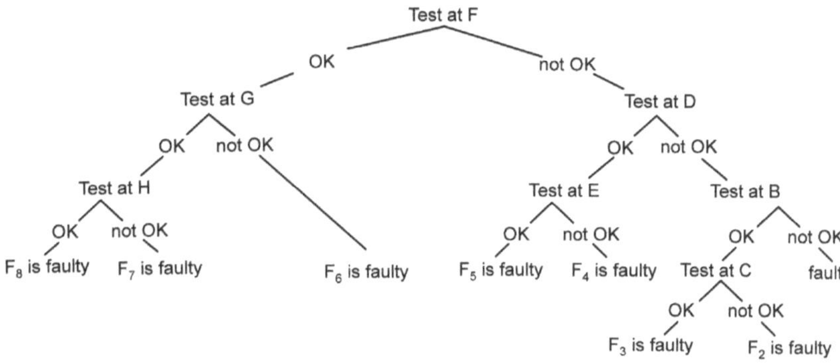

FIGURE 5.17 Decision tree for the midway check method for the above situation.

5.24 Summary of the features of management as a system

All the points described in the previous paragraphs can be summarized as below (under)

1. **Management is a system:** Management can be defined as a system and unlike the biological and mechanical systems, it has the characteristics of a social system. It has subsystems which are integrated as a whole.
2. **Management is an open system:** It has an environment and with interaction, management takes its resources or inputs, allocates and combines the resources to produce the desired output to the environment.
3. **Management is dynamic:** The equilibrium of the organization is ever changing, moving towards the growth and expansion by preserving some of its energy. Thus, the organizational effectiveness is determined by the ratio of energy exchanges.
4. **Management is probabilistic:** A deterministic model always specifies the use of a situation in condition with predetermined results. Management being probabilistic, points out only the probability and never the certainty of performance and the consequent results. Nevertheless, for aiding decision making, the management assumes a certain amount of deterministic models to arrive at some conclusions and then do the analysis in combination with the probabilistic approach.
5. **Management is multi-dimensional:** System approach points out complex multilevel and multi directional characters. At the macro level, management can be applied to the business systems as a whole. At the micro level, it can be applied to an organized unit or any if its elements.

6. **Management has multiple variables:** There is no simple cause-effect phenomenon like in physics. An event may be a result of so many variables which themselves are interrelated and interdependent.
7. **Management is adaptive:** The survival and growth of an organization in a dynamic environment demands an adaptive system that continuously adjusts to changing environment through feedback mechanism provides information to take corrective action for achieving the desired results.
8. **Management is multi-disciplinary:** management being a system draws integrated knowledges from several disciplines and schools of thought. In fact, integration of the relevant aspects of various disciplines is the real contribution to the management as a system.

5.25 Conclusion

The chief purpose of the systems approach is to provide the management with an analytical framework by which it can identify, describe and interrelate the process and components that make up a particular system. In other words, the systems approach enables a manager to maintain a perspective of the whole process while he analyzes the parts.

Criteria questions

1. Explain the concept of 'Productive System'. (5.3)
2. There are several concepts of not contractions involved in defining the term System. Discuss. (5.3)
3. Explain the 3 basic elements of a system. (5.4)
4. Illustrate the educational system. (5.5)
5. Why is input control important? (5.5)
6. Distinguish between open and closed systems with illustrations. (5.8)
7. What are subsystems and how are they related to the systems. (5.9)
8. Discuss the components of a management cube. (5.12)
9. What do you understand by planning pyramid? Illustrate with an organizational situation. (5.13)
10. Discuss the types of information required for optimal planning. (5.15)
11. Is management a system? Justify. (5.16)
12. Distinguish between the terms decide, determine, settle and conclude. (5.14)
13. Discuss the characteristics of decision making. (5.16)
14. Distinguish between decision making under risk and decision making under conflicts. (5.17)
15. Discuss the different approaches to decision making. (5.19)
16. Represent graphically the systematic decision-making process. (5.19)
17. Is group thinking good or bad for decision making? Discuss. (5.22)
18. What do you understand by Decision Tree? (5.23)

Further reading

1. Chenhall RH, Langfield-Smith K. *Accounting, organizations and society*. Elsevier; 1998.
2. Kiran DR. *Maintenance engineering and management: precepts and practices*. BS Publications; 2014.
3. Kiran DR. *Total quality management: an integrated approach*. BS Publications; 2016.
4. Kiran DR. *Production planning and control, a comprehensive approach*. BS Publications; 2017.
5. Jenkins GM. *Systems approach to management*. 1968.
6. www.yourarticlelibrary.com/management.
7. en.wikipedia.org/wiki/Approaches_of_management.
8. study.com/.../systems-approach-to-management-theory-lesson-quiz.html.

Chapter 6

Method study – select

6.1 Introduction

In the previous chapter, we have understood the basic procedure of work study and also learnt that its two wings are method study and work measurement. In the following chapters, we will study method study in detail. As we have already seen method study is defined as the systematic recording and critical examination of the existing and proposed ways of doing a work, as a means of developing and applying easier and more effective methods of reducing the costs.

6.2 Concept of method study

Method study is basically conducted to simplify the work or working yielding higher productivity. It is always desirable to perform the requisite function with desired goal of minimum consumption of resources. The defined method signifies how a work is to be done, i.e. description of how we consume resources in order to achieve our target.

Methods are integral part of work accomplishment and signify,

1. How well our methods utilize the limited available resources such as manpower, machines, materials and money.
2. How our methods physically affect the production output of the unit.
3. The quality of output obtained by application of our methods.

Thus, the methods adapted can determine the amount of input materials, time, power and money consumed and hence form the core where one can attempt to reduce the consumption of resources so as to reduce the cost per unit output through utilization of proper methods. The method design can decide the cost and quality of output produced.

6.3 Definitions of method study

As seen above method study is a procedure for examining the various activities associated with the problem which ensures a systematic, objective and critical evaluation of the existing factors and in addition and imaginative approach while developing improvements.

Work Organization and Methods Engineering for Productivity.
DOI: https://doi.org/10.1016/B978-0-12-819956-5.00006-6

As defined in the previous chapter, we can repeat the definitions on method study

Method study is the systematic recording and critical examination of the existing and proposed ways of doing a work, as a means of developing and applying easier and more effective methods of operations, thereby reducing costs.

I.L.O.

Method study is subjecting each part of a given piece of work to close analysis to eliminate every unnecessary element or operation, as a means of approaching the quickest and best method of performing the work. It also includes formulation of incentive schemes, and improvement and standardization of equipment, methods, operator training, working conditions, etc. Also called methods engineering.

Business dictionary.

Method study is the process of subjecting work to systematic, critical scrutiny to make it more effective and/or more efficient. It is one of the keys to achieving productivity improvement.

Institute of Management Services.

6.4 Scope of method study

The task of method study for work simplification and work system design covers the followings:

1. Layout of shop floor and working areas or work stations.
2. Working conditions, i.e. ergonomics, etc.
3. Handling distances (material movement)
4. Tooling and equipment used.
5. Quality standards to be achieved.
6. Operators and operations in achieving the production targets.
7. Materials to be used.
8. Power required and available.
9. Work cycle time.
10. Working processes.

6.5 Aims of method study

- Optimum output.
- Improved utilization of resources.
- Improved flow of work.

- Efficient processes and procedures.
- Effective materials handling.
- Improved layout of factories and offices.
- Better design of plant, equipment and buildings.
- Better working conditions and environment.
- Higher standards of safety and health.
- Higher earnings and greater job satisfaction.
- Waste reduction.
- Optimum inventory level.
- Standardization and rationalization.
- Better quality
- Improved administration
- Higher returns on investment
- Increased market, and customer satisfaction

6.6 The three levels of method study

We had seen in the last chapter that method study encompasses motion study besides operational analysis; we can say in general that method study is conducted in three levels.

a. Method study proper, which is the broad investigation and improvement of a total section or group of operations or improvement of a single operation in relation with the other operation. This generally involves operations long enough to be studied by stopwatch or normal eye observations.

b. Motion study, which is a more detailed study of the individual operator emphasizing on the improvement of the body motions of the operator as he performs the operation, with relation to the layout, position of tool, etc. environment, posture, etc. In general, motion study is the analyzes of the hands, legs and eye movements occurring during an operation or work cycle for the purpose of eliminating wasted movements and establishing a better sequence and co-ordination of movements.

c. Micro motion study, which is the study of still finer elements which cannot be normally observed by naked eye, and which requires the help of motion cameras for analysis. This analysis is characterized by the use of therbligs, about which we will know more in later chapters.

6.7 The basic procedure for method study

The basic procedure for method study is better understood by the word "SREDDIM".

Select	the job to be studied.
Record	all the factors about the present method of operation in sufficient details, by use of easily communicable representations like the charts & diagrams.

Examine	these factors critically separating the unproductive and foreign elements from the productive elements and also question the original sequence using techniques best suited for this purpose.
Develop	alternative methods which are practical, economic and effective, having due regards to all the factors,
Evaluate	the above alternatives and select the optimum among them.
Define	the new method so that it can always be identified and interpreted correctly by anyone, anytime.
Install	the new method as a standard practice, and
Maintain	the standard practice by regular routine checkups.

Fig. 2.3 of Chapter 2 may be referred to for the graphical representation of the procedure. We have in this paragraph listed the steps one by one. As we would be seeing in the chapter on creativity, this order is not rigid.

ILO in its fourth edition of Introduction to Work Study, adds a fifth step 'Evaluate' after Developing, highlighting the importance of evaluating the various alternatives developed and then selecting the best among them. However, in the subsequent paragraphs and chapters of this book, the evaluation is treated as a part of development.

6.8 Method study - selection of the jobs

The first step in method study is to select a suitable job and define the objectives of the study. This is generally not as simple as it appears to be. The tricky part often lies in identifying the real problem as distinct from the apparent problem. As seen in paragraph 1.11, the line managers or supervisors, despite being more experienced in the job, are preoccupied more with their own routine problems and so cannot identify the real problem for the method study. The most common example being the line manager complaining of the shortage of hands for a particular operation or a group of operations resulting in a bottleneck, and requesting the work study engineer to assess the additional hands needed for smooth running. This is the apparent problem. On investigation, it may turn out that the operation itself has a large content of ineffective time and the real problem is thus of ineffective job design.

Hence the problem should be formulated in such a manner so as to include as much as the total problem or the real problem, as the economics of the situation and organizational boundaries permit. In other words, the problem should be stated in such a way to define the root problem clearly.

6.9 Factors involved in the selection of jobs for method study

While selecting a job for method study, the following considerations must be borne in mind in order to achieve acceptance from all concerned.

1. **Human considerations:** They play a very critical role in the successful implementation of their commendation. Any change in the job or method always results in a certain amount of resistance either from supervisors or workers as discussed further in Chapter 20 The emotional reactions to the investigation and method change should be anticipated and co-operation should be sought from the concerned persons. They should be freely involved in discussions throughout the study. Further reference may be made to the discussions as well as the case study detailed in Chapter 1.

   An intelligent way of selecting jobs for method study would be to assign priority for the jobs that are known to be unpleasant or over fatiguing. The operators naturally start taking more interest and give more co-operation in the data collection. If on implementing, the working conditions are really bettered, then the method study man would indeed have won the confidence of the shop floor personnel aiding their acceptance.
2. **Economic considerations:** Since any method study basically aims at the overall cost reduction, these considerations form the basis for a majority of the investigations. It is an obvious waste of time and energy to start or continue a long investigation if the economic importance or the total cost saving achieved by changing the method is insignificant or when the job is not expected to run long, There is no point in a highly paid engineer to spend two months on a project that results in a cost saving of few hundred rupees. We should always ask the question.

   "Will it pay to do method study on this job?"

   "Is it justified to continue the study?"

   In other words, one should always be on the lookout for jobs involving large ineffective work contents. The most common examples are:
   a. Bottlenecks that not only incur high production costs, but also hold up subsequent production operations.
   b. Excessive materials handling, especially those involving manual handling,
   c. Excessive scrap generating operations,
   d. Poor utilization of resources, and
   e. Repetitive operations.
3. **Technical consideration:** Before selecting the job, we should ensure that the analysis and implementing would be technically feasible. That is,
   a. Adequate technical knowledge should be available to carry out the analysis. For example, if a method study project is undertaken on cloth weaving operation, the analyzer must have a good basic knowledge of the weaving process, the various quality considerations, as well as the types, capacities and limitation of the process machinery not only available with the company but elsewhere. At least he should be in a position to get continuous help from the process specialists.

b. The recommended process method itself should be feasible. There is no point in recommending for higher speeds or feeds if the machine is incapable of such speeds due to its condition, or recommending shop to shop handling by fork lift truck, if the gangways are not wide enough.

6.10 Conclusion

The following tips on job selection would help an easier winning of acceptance for the proposed improvements.

1. Select jobs that are highly fatigue creating, uncomfortable or generally considered as dirty by the operators, like heavy manual work, bad working conditions, etc.
2. Give preference to the jobs requested by the line manager.
3. Bottleneck jobs that are generally recognized by all as causing regular production bottlenecks.
4. Jobs that involve high production costs.
5. Jobs that involve excessive scrap generation.
6. Jobs where the collection of all required data is not impossible.
7. Ensure that the work study man has adequate or basic knowledge, on the process in which study is made.
8. Avoid choosing small bench jobs or non repetitive jobs which require considerable detailed study, but yield only low savings compared to repetitive jobs.
9. In most cases you can visualize the likely improved method in the first instance itself. In such cases do not proceed unless you are confident that you can overcome the stiff resistance if any, for your recommendation. Rehearse yourself to face such a situation.

Criteria questions

1. What is method study and what activities does it signify? (6.2)
2. Discuss the three levels of method study. (6.6)
3. What is SREDDIM? Explain each letter (6.7)
4. What are the 3 Factors involved in the selection of jobs for method study? Explain. (6.9)

Further reading

1. ILO. *Introduction to work-study.* 3rd ed. I.L.O. 1979.
2. Shaw A. *The purpose and practice of motion study.* Columbia Press;1960.
3. Kiran DR. Production planning and control: a comprehensive approach. BS Publications; 2017.

Appendix

Checklist of factors for selection, examination and development of jobs

I.L.O. in its book Introduction to Work-study, 4th edition (1992) has given a comprehensive checklist for the selection of jobs and development of alternatives, as given in the appendix below.

Reference can also be made to The Purpose and Practice of Motion Study, by Anne Shaw, for a sample checklist of the various factors that should be considered before selecting a project for method study.

Source: *While most points are from ILO book, some from Anne Shaw's book also are added. This author wishes to acknowledge his thanks to both these books.*

Checklist for examining and developing a new method of work

A. Operations
B. Products and parts design
C. Quality requirements
D. Materials utilized (utilization)
E. Workplace layout
F. Materials handling
G. Work organization
H. Working conditions
I. Job enrichment

A. Operations

1. What is the purpose of the operation?
2. Is the result obtained by the operation necessary? If so, what makes it necessary?
3. Is the operation necessary because the previous operation was not performed correctly?
4. Is the operation instituted to correct a condition that has now been corrected otherwise?
5. If the operation is being carried out to improve appearance, does the additional cost give extra stability?
6. Can the purpose of the operation be obtained in another way?
7. Is the operation being performed to satisfy the requirements of all users of the product, or is it made necessary by the requirements of one or two customers only?
8. Does a subsequent operation eliminate the necessity for this operation?
9. Was the operation established to reduce the cost of a previous operation, or a subsequent operation?
10. Would adding a further operation make other operations easier to perform?
11. Is there another way to perform the operation and maintain the same or even better results?
12. Have conditions changed since the operation was added to the process?

13. Could the operation be combined with a previous or a subsequent operation?
14. Can the operation analyzed be combined with another operation? Can it be eliminated?
15. Can it be broken up and the various parts of the operation added to other operations?
16. Can a part of the operation being performed be completed more effectively as a separate operation?
17. Is the sequence of operations the best possible or would change the sequence improve the operation?
18. Could the operation be done in another department to save the cost of handling?
19. If the operation is changed, what effect will it have on the other operations? On the finished product?
20. If a different method of producing the part can be used, will it justify all the work and activity involved?
21. Can the operation and inspection be combined?

B. Products and parts design

1. Can the design be changed to simplify or eliminate the operation?
2. Can the number of component parts be reduced?
3. Can certain component parts be standardized?
4. Can a standard part be substituted by another cheaper or better material?
5. Has Pareto analysis been used to detect the products or parts that are most valuable?

C. Quality requirements

1. Has an agreement has (has added) been reached by all concerned as to what constitutes acceptable quality?
2. What are the inspection requirements for this operation?
3. Can the operative inspect his or her own work?
4. Are tolerance and other standards appropriate?
5. Can standards be raised to improve quality without unnecessary cost?
6. Will lowering standards reduce costs considerably?
7. Can the finished quality of the product be improved in any way above the present standard?
8. Can the quality be improved by using new processes?
9. Are the same standards necessary for all customers? 1
10. Will change in standards and inspection requirements increase or decrease the defective work and expense in the operation, shop or field?
11. What are the main causes of rejections for the part?
12. Would a change in the composition of a product render it less susceptible to quality variations?

D. Materials utilized (utilization)

1. Is the material being used really suitable for the job?
2. Could a less expensive material be substituted and still do the job?
3. Could a lighter-gauge material be used?
4. Is the material purchased in a condition suitable for use?
5. Could the supplier perform additional work on the material that would improve usage and decrease waste?
6. Is the material sufficiently clean?
7. Is the material bought in amounts and sizes that give the greatest utilization and limit scrap, effects (offcuts) and short ends?
8. Is the material used to the best possible advantage during cutting, processing?
9. Are indirect materials used in connection with the process - oils, water, acids, paint, gas, compressed air, electricity - suitable, and is their use controlled and economized?
10. How does the cost of material compare with the cost of labor?
11. Can the design be changed to eliminate excessive loss and scrap material?
12. Can the number of materials used be reduced by standardization?
13. Can the part be made from scrap material or offcuts?
14. Can the scrap be salvaged for further processing?
15. Can the scrap be sorted out for sales at higher price?
16. Is the supplier of the material performing operations on it which are not necessary for the process?
17. Is the material supplied of consistent quality?
18. Could a more careful inspection of incoming material decrease difficulties now being encountered in processing?
19. Is the material free from sharp edges and burrs?
20. What effect does storage have on material?
21. Could sampling inspection combined with supplier rating reduce inspection costs and delays?
22. Could the part be made more economically from off cuts in some other gauge of material?

E. Workplace layout

1. Does the plant layout aid efficient material handling?
2. Does the plant layout allow efficient maintenance?
3. Does the plant layout provide adequate safety?
4. Is the plant layout suitable for appropriate sequencing of operation? Can parts of an intermittent operation be changed to a line operation layout for major products or parts?
5. Does the plant layout help social interaction between the operatives?
6. Are materials conveniently placed at the workplace?
7. Are tools pre-positioned to save mental delay?

8. Are suitable jigs and fixtures available at the workplace to facilitate work, particularly in assembly operations?
9. Are adequate working surfaces provided for subsidiary operations, e.g. inspection and deburring?
10. Are facilities provided for the removal and storage of scrap and waste?
11. Is adequate provision made for the comfort of the operative, e.g. fan, duckboard or chairs?
12. Is the lighting adequate for the job?
13. Has provision been made for the storage of tools and gauges?
14. Has provision been made for the storage of the operatives' personal belongings?

F. **Materials handling**

1. Is the time spent in bringing material to the workstation and in removing it large in proportion to the time used to handle it at the workstation?
2. If not, could material handling be done by the operatives to provide a rest through change of occupation?
3. Should hand, electric or fork-lift trucks, or conveyors or chutes be used?
4. Should special racks, containers or pallets be designed to permit the handling of material with ease and without damage?
5. Where should incoming and outgoing materials be located in the work area?
6. Can material be dispatched from a central point by means of a conveyor?
7. Is the size of the container suitable for the amount of material transported?
8. Can a container be designed to make material more accessible?
9. Could a container be placed at the workstation without removing the material?
10. If an overhead traveling crane is used, is the service prompt and accurate?
11. Can gravity be utilized by starting the first operation at a higher level, and using suitable chutes or conveyors?
12. Are truck loading and unloading stations located appropriately?
13. Would a turntable eliminate walking?
14. Can incoming raw material be delivered at the first workstation to save double handling?
15. Could operations be combined at one workstation to save double handling?
16. Would a container of standard size eliminate weighing?
17. Are containers uniform to permit stacking and eliminate excessive use of floor space?

18. Could material be bought in a more convenient size for handling?
19. Would signals, i.e. lights, bells, etc., notifying workers that more material is required, save delay?
20. Can the location of stores and stockpiles be altered to reduce handling and transport?

G. Work organization

1. How is the job assigned to the operative?
2. Are things so well controlled that the operative is never without a job to do?
3. How is the operative given instructions?
4. How is material obtained?
5. How are drawings and tools issued?
6. Is there a control on time? If so, how are the starting and finishing times of the job checked?
7. Are there many possibilities for delays at the drawing-room, toolroom and storeroom?
8. Is the material properly positioned?
9. If the operation is being performed continuously, how much time is wasted at the start and end of the shift by preliminary operations and cleaning up?
10. What clerical work is required from operatives for filling in time cards, material requisitions and the like? Can some of these operations be computerized?
11. How is defective work handled?
12. How is the issue and servicing of tools organized?
13. Are adequate records kept on the performance of operatives?
14. Are new employees properly introduced to their surroundings and do they receive sufficient instruction?
15. When workers do not reach a standard of performance, are the details investigated?
16. Are suggestions from workers encouraged?
17. Do the workers really understand the incentive plan under which they work?

H. Working conditions

1. Is the lighting, even and sufficient at all times?
2. Has glare been eliminated from the workplace?
3. Is the proper temperature for comfort provided at all times? If not, can fans or heaters be used?
4. Would installation of air-conditioning equipment be justified?
5. Can noise levels be reduced?
6. Can fumes, smoke and dirt be removed by exhaust systems?
7. If concrete floors are used, are duckboards or matting provided to make standing more comfortable?
8. Can a chair be provided?

9. Are drinking fountains with cool water provided and are they located nearby?
10. Has due consideration been given to safety factors?
11. Is the floor safe, smooth but not slippery?
12. Has the operative been taught to work safely?
13. Is the clothing suitable from a safety standpoint?
14. Does the plant present a neat and orderly appearance at all times?
15. How thoroughly is the workplace cleaned?
16. Is the plant unduly cold in winter, or stuffy in summer, especially on the first morning of the week?
17. Are dangerous processes adequately guarded?

I. Job enrichment

1. Is the job boring or monotonous?
2. Can the operation be made more interesting?
3. Can the operation be combined with previous or subsequent operations to enlarge it?
4. What is the cycle time?
5. Can the operative do his or her own setting?
6. Can the operative do his or her own inspection?
7. Can the operative deburr his or her own work?
8. Can the operative service his or her own tools?
9. Can the operative be given a batch of tasks and do his or her own scheduling?
10. Can the operative make the complete part?
11. Is job rotation possible and desirable?
12. Can group work be encouraged?
13. Are flexible working hours possible and desirable?
14. Can buffer stock be provided to allow variations in work place?
15. Does the operative receive regular information about his or her performance?

Chapter 7

Method study – record

7.1 Data collection

For proceeding with any study, collecting the required data forms the first step. Data collection may be one or more of

- Direct observation
- Interviewing the concerned persons and
- Examination of the relevant documents and records.

The data relevant to the study is generally grouped into

a. Background information like the organization details of the company, function and tasks, and availability of capital.
b. Details of the product, the design specifications, its assemblies and subassemblies.
c. Process details and operational details.
d. The materials handling/transporting involved. Though this is generally a part of the operation, emphasis is made here since in a majority of the cases, the nonproductive component of handling contributes largely to the ineffective time.
e. The physical means of executing the operation.

7.2 Symbols and charts

Why symbols and charts? The usual way to record an operation is of course by writing it down in detail. This is not only difficult and time consuming, but the critical points are differently interpreted by different persons. Take for example the information given in the following paragraph.

A survey of adult males in a town, taken in September 1984, by personal interview, showed that 122 of the 2049 single men under 30 years attended the cinema less than once a month, 1046 attended one to four times a month, and 881 more than four times a month, Of the single men 30 and over the respective figures were 374, 202 and 23, a total of 599. As regards television viewing, 830 of the single men under 30 viewed less than 25 hours a week, the other 1219 viewing 15 hours or more. For single men 30 and over these figures were 358 and 241 respectively. As regards the married men, 1404 of the 'under 30

Work Organization and Methods Engineering for Productivity.
DOI: https://doi.org/10.1016/B978-0-12-819956-5.00007-8

years' group attended the cinema less than once a month, 289 attended one to four times a month and 112 more than four times month. For those of 30 and over the figures were 1880, 115 and 10 respectively. TV viewing figures showed that 1162 married men under 30 viewed less than 15 hours a week and the remaining 643, 25 hours or more. Of the 'over 30 years' group, 484 viewed less than 15 hours and 1521 viewed 15 or more hours a week.

As obvious from the reading of the above data, this method is monotonous and a very ineffective method of data presentation. As the figures are jumbled up amongst words, the mind does not register them well enough to retain in memory.

7.3 Tabular presentation

Now let us how the same data looks in a table form (Table 7.1). A look at the left of the table gives you a picture of the several categories and then a glance of the right half enables the mind to immediately link the figures with the corresponding category and thus register better in the memory.

While the tables give a concise presentation of statistical data, it cannot effectively present operational data. In work study, practice we prefer using charts and diagrams over the tables for the visual presentation of the operational data. Consider another example of the operational sequence of a certain product narrated in the following paragraph.

TABLE 7.1 Tabular representation of the data.

Category	Frequency	Single		Married	
		Under 30	Over 30	Under 30	Over 30
Cinema viewing	Less than once a month	122	374	1404	1880
	1 to 4 times a month	1046	202	289	115
	More than 4 times a month	881	23	112	10
	Subtotal	**2049**	**599**	**1805**	**2005**
Television viewing	Less than 15 h a week	830	358	1162	484
	More than 15 h a week	1219	241	643	1521
	Subtotal	**2049**	**599**	**1805**	**2005**

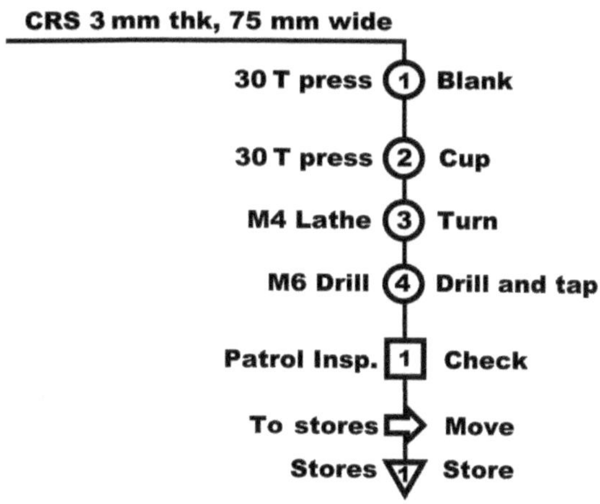

FIGURE 7.1 A chart with symbols.

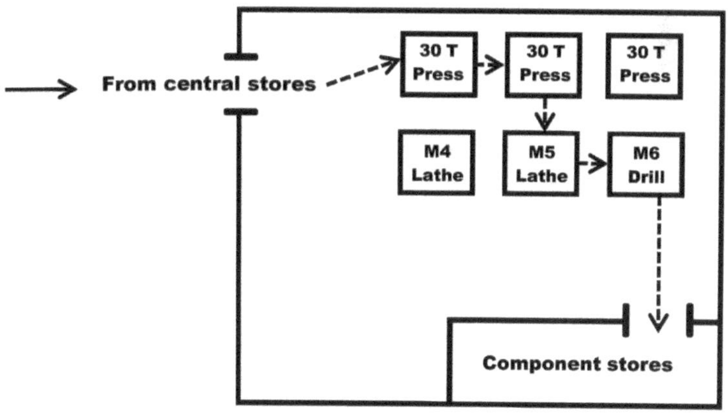

FIGURE 7.2 A flow diagram.

A bell cup of a certain bicycle component is manufactured by first blanking from a 3 mm thick, half hard CRS strip on a 30 T press, then turning it on a lathe and then drilling and tapping on a drilling machine. It is then checked for quality before sending it to the stores.

This description may not give a clear picture of what and where the item is worked upon. But if the same is represented in a chart form as per Fig. 7.1, the picture would be clear.

In addition, if we represent these movements on a plan of the shop, may or may not be drawn to scale as in Fig. 7.2, the picture will still be clearer.

The significance of the charts and symbols will be clear from Figs. 7.9 and 7.10, which illustrate a two hand process chart, where both hands are

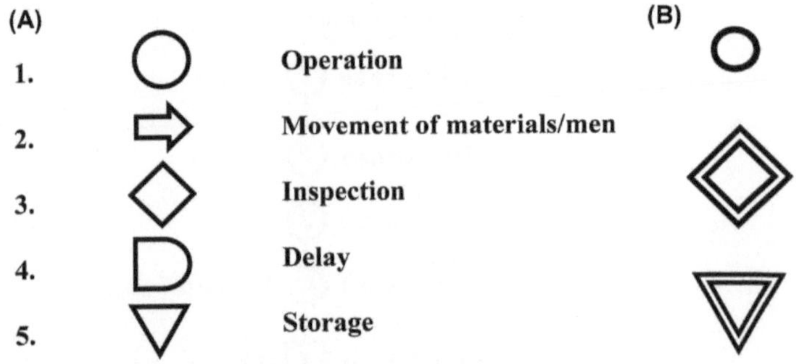

FIGURE 7.3 Illustration of process chart symbols.

working simultaneously. It is imperative that every line, figure, picture or symbol in the diagram or chart should be understood or interpreted in the same way by everyone. The charts and symbols are hence developed to international standards so that, each symbol conveys the same meaning,

7.4 Symbols

Symbols are used for recording the nature of events. In work study, there are generally three groups of symbols, the first being the process chart symbols, used in a majority of method study projects, the second being the therbligs, generally used in motion study. The third group consists of more detailed symbols used in computer flow charts.

7.4.1 Process chart symbols

In general 5 symbols as illustrated in Fig. 7.3 are used as the process chart symbols.

7.4.2 Some variations in the process chart symbols

While the above are the internationally accepted symbols mostly recommended for method study, some variations have been adapted by some practitioners for the sake of simplicity.

A circle smaller than the operation symbol, used by Gilbreth etc. for transportation. As long as the size of the circle is distinctly smaller than that of the operation, this is easier to draw than the standard symbol, the arrow.

Inspection of quantity while the usual single line square is used for the inspection of quality.

For finished good storage, as distinct from single line triangle which is used only for the in-process storage of semi-finished goods. This variation is especially useful in the preparation of the outline process charts illustrated in Fig. 7.5.

7.4.3 Therbligs

While process chart symbols are used for the method study techniques, therbligs are used for the micro motion analysis, which is the study of the finer motions of the body movement in a fixed workplace. The term therblig is coined as an anagram of its initiator, Gilbreth, one of the pioneers of work study, referred in Chapter 2. These are dealt in more detail in the Chapter 13 on Principles of Motion Economy, and illustrated in Fig. 13.3.

7.5 Charts used in work study

Recording is the very basis for critical examination. Charts provide a very popular method of recording. As we have seen earlier, what needs ten pages to describe a process verbally needs just one page of symbols and figures arranged in a graphical form providing a better visual concept. The charts used in work study basically fall into three groups.

7.5.1 Charts indicating the process sequence

Also called as process charts, they help in visualizing the process by representing the sequence of operations by use of symbols.

1. Outline process charts
    - Flow process charts
    - Man type
    - Material type
2. Equipment type
3. Two handed process charts
4. Flow charts
5. Process charts.

7.5.2 Charts using a time scale

1. Multiple activity charts
2. Simultaneous motion charts (SIMO Charts)
3. Therblig charts
4. PMTS charts
5. Bar charts.

7.5.3 Diagrams indicating movements

1. Flow diagrams or route diagrams
2. String diagrams
3. Travel charts
4. Cyclographs

5. Chronocyclographs
6. Memo Motion Photography
7. Time-Lapse camera vedio

Each one of these charts is further explained in the following paragraphs. It may be brought to the notice of the reader here that the nomenclatures used for some of the charts vary from book to book. Most European authors, including ILO refer to the chart per para 7.6 as outline process chart whereas American authors, including HB Maynard in his Industrial Engineering Handbook refer to this as operations process chart. Both names are valid and can be treated as synonyms. However, to avoid such confusion, this book sticks to the nomenclature used in the ILO book.

7.6 Outline process charts

> *The outline process chart is a chart that gives a bird eye's view of the various operations, inspections and storage done in sequence for all the components that go into a particular product or assembly.*
>
> B.S. 3138.

Several components that go into a single assembly or subassembly can be represented in a single outline process chart. Each component is identified at the top on a horizontal line, by the name and its raw material specification in brief as shown in Fig. 7.4. The number of such components that go into one final product for which the OPC is drawn, is indicated as no. off. All the operations, inspections and the storage activities are indicated along a vertical line by use of symbols, as illustrated in Fig. 7.5, are written on the right side of the line while the machine (by name or by code) is written on the left side of the line. Sometimes the standard time of the operation is indicated either inside the circle of the operation symbol or to its left.

The line for the major component is drawn in this manner at the extreme right, while those for the other components that are assembled to the first by bolting, welding, etc., join this line from the left at the appropriate place. Fig. 7.5 illustrates how to translate a lengthy process description of an assembly used in the bicycle manufacture as detailed below, into an outline process chart.

Example: The following three components are used in the assembly of a product.

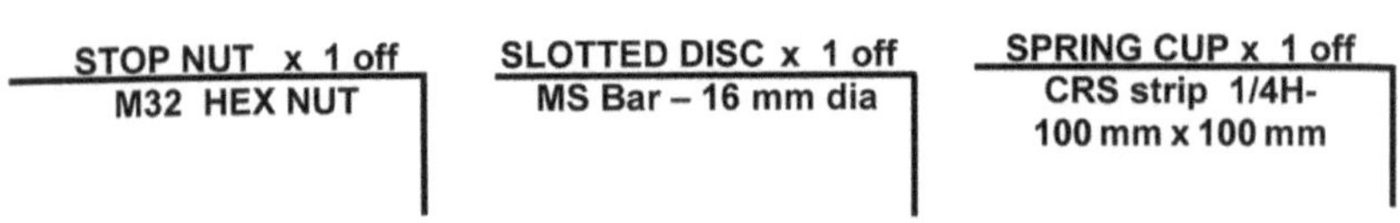

FIGURE 7.4 Upper portion of the operation process chart.

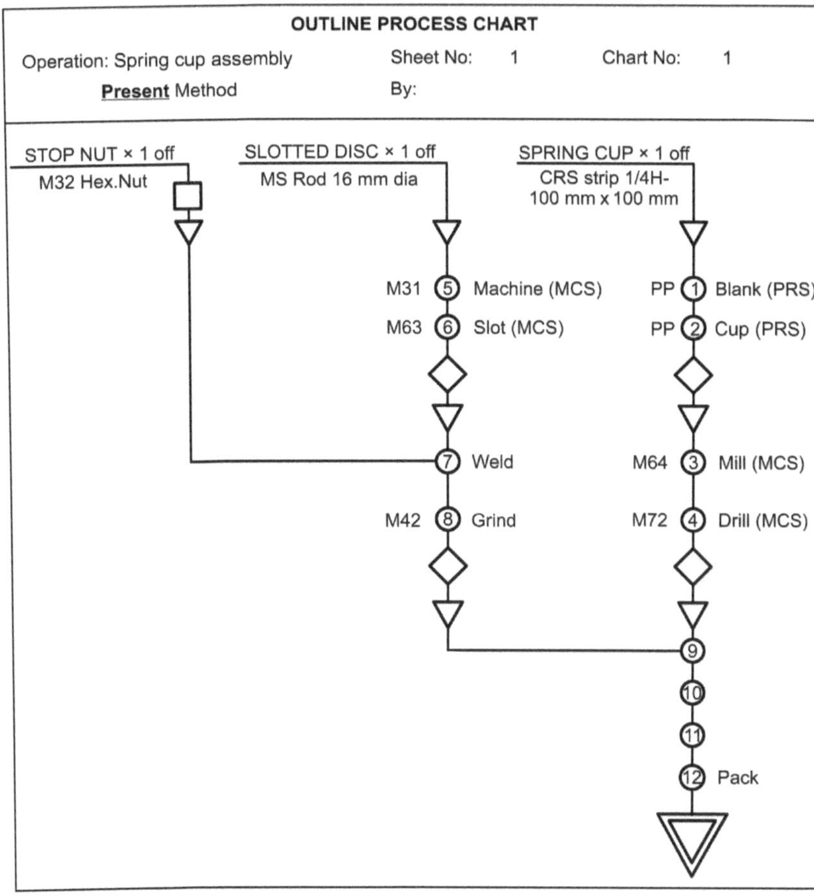

FIGURE 7.5 Outline process chart for the Subassembly.

Component A: *Receive as raw material which is cold rolled steel strip, ¼ hard, 100 mm wide x 3.0 mm thick - store - blank at press shop on a 50 ton press - cup on a 32 ton press - mill in the machine shop on machine no. M64 - drill on machine no. M72 - assemble with the subassembly of components B&C at the assembly shop - polish - finish pack - finish store.*

Component B: *Receive as raw material which is MS round bar, 16 mm diameter - machine on auto lathe no. M31 in the machine shop - slot on machine no. M63 - weld with component C at the welding shop - grind at the machine shop on machine no. M42 - store at production stores before moving to the assembly shop for assembling with the component A.*

Component C: *Purchase as M12 Hexagonal nut - weld to component B after due inspection and storage.*

Question: Build up an outline process chart for this product. Add inspections and storage where you feel necessary.

7.7 Flow process chart

Flow process can be defined as follows

Flow process chart sets out the sequence of the flow of a product or an equipment or a man, by recording all the events under review using appropriate process symbols. These flow process charts can either be material type or equipment type or man type, depending whether the subject being charted is the material, equipment or man.

As the definition indicates, the flow process charts can be of three types.

a. Material types, per Fig. 7.6, wherein the flow of the materials or the product or the components are charted. A majority of the flow process charts are of this type.

b. Equipment type, wherein the movement of certain equipment like welding equipment, portable drills, cranes, fork lift trucks and air compressors, that are taken from work place to workspace on a regular basis. This chart would be useful if this equipment causes bottlenecks or excessive waiting time at the workplace that needs it. For example, the bus in Fig. 7.7 needs to be gas welded at 4 places A to D, and only one gas welding equipment is available. The flow process chart for this can be drawn as per Fig. 7.8.

		FLOW PROCESS CHART			Material type						
Dept.	Opn. no	Operation: Ore loaing									
		Element	Qty. tonnes	Dist. m	Time m	Symbol ○	➡	D	□	▽	Remarks
	1	Arrive in wagons	22 ton		0.00		●				
	2	Open wagon doors			0.50	●					
	3	Unload ore			23.00	●					Improve
	4	Clear track			12.50	●					Eliminate
	5	Move wagon			2.00		●				
	6	Position next wagon			1.00						
		Total time			39.00	3	2				

FIGURE 7.6 Material type flow process chart.

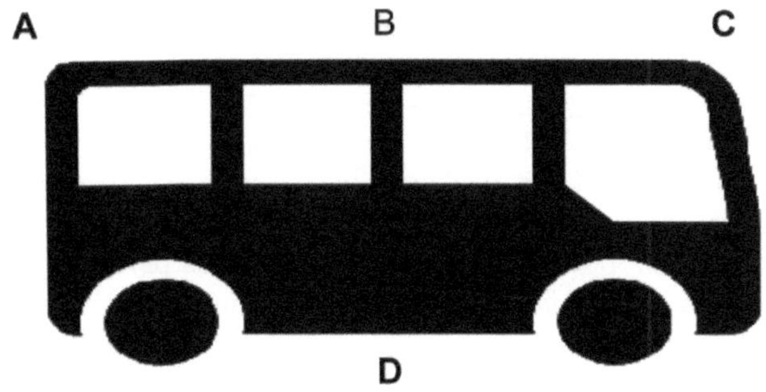

FIGURE 7.7 Gas welding to be done on a bus.

	FLOW PROCESS CHART				Equipment type					
Opn. no	Operation: Gas welding									
	Element	Qty. tonnes	Dist. m	Time m	Symbol ○	➡	D	□	▽	Remarks
1	Move to A		20	5.00		●				
2	Gas weld			6.00	●					
3	Inspect			1.00				●		
4	Move to B		3	1.50		●				
5	Gas weld			10.00	●					Improve
6	Inspect			1.00				●		
7	Move to C		3	1.00		●				
8	Wait			5.00			●			Eliminate
9	Gas weld			12.00	●					Improve
10	Inspect			2.00				●		Improve
11	Move to D		6	1.00		●				
12	Gas weld			8.00	●					Improve
13	Inspect			2.00				●		Improve
14	Move to stores		20	5.00		●				
	Total				4	5	1	4		

FIGURE 7.8 Equipment type flow process chart.

c. Man type, wherein the activities of a single worker or a gang moves from place to place as a part of their work, like the maintenance gang or the road laying gang, this type of chart would be used.

Flow process charts are specifically useful in the following 3 situations

- The work sequence or elements are not exactly identical, but vary from cycle to cycle. This variation may be due to the operator's practices like cleaning an equipment or may be due to the inherent nature of the work like the maintenance operations or like the powdering of the hardened fertilizer in bulk storage.

- The work is not cyclic but unique like the port workforce handling different types and sizes of the packages in clearing a general cargo storage area.
- The work is cyclic but includes several sub-cycles performed with different frequencies. A typical example is in packing operations, where small automobile components are packed in individual cartons, then 10 of them are packed in a larger carton and 4 or 6 of the larger cartons packed in wooden boxes, all these forming one cycle. In this case it is essential that to indicate the frequency of each element and the sub-cycle.

The format used is generally common, whatever the type of chart, as illustrated by Fig. 7.6, bearing in the heading, man/material/equipment. While filling up the form, the two words not needed would be canceled. In some cases, a simpler form of a flow process chart is drawn, as in Fig. 7.8, though not much preferred. Nevertheless, this can be used for an initial recording, which may later be transferred to the chart of the former format.

The process chart can be a valuable future reference for the following information as provided on the chart:

1. The name of the product, material or equipment charted, with drawing numbers or code numbers.
2. The job or process being carried out, clearly stating the starting and end activities, and whether the method is the present or proposed.
3. The location in which the activity is taking place (department factory, site, etc.)
4. The chart reference number, the sheet number and the total number of sheets.
5. The observer's name and if necessary, the name of the approval,
6. The date of the study,
7. A key to the symbols used,
8. A comparison of the number of occurrence of each of the activity like operation, movement, etc., in the present as well as the proposed methods and
9. A summary of the distance, time and if necessary and possible, the cost of labor, material, etc., as a comparison of the two methods.

7.8 Differences between the outline process chart and the flow process chart

(a) Outline process chart gives the details of all the components in a product or assembly, whereas flow process chart gives the details of one component only (if it is of material type). Thus, the former gives a clear bird's eye view for the whole process, especially how various components are assembled together.

(b) Outline process chart does not indicate movements or delays, which are the basic nonproductive elements, while the latter indicates them specifying the distance moved and time spent in each case. Thus, a short glance of the chart is enough to draw our attention to the extent of these nonproductive elements.

(c) An outline process chart can be drawn only to the components/products that is suitable for material flow, whereas the later can be drawn for the flow of an equipment. (For example, when a portable drill is used for a variety of operations by various persons, and a study is to be done to effectively utilize this item which is in high demand), or of a man (for example, the maintenance gang).

(d) The flow process chart is especially useful for charting bulk material flow, wherein in each operation, it is handled in varying quantities, like handling of iron ore from the train to the ship (case study in Chapter 10) or testing, moving and packing of batteries.

(e) Flow process chart should not be confused with a flow chart, which is generally used in process industries as detailed in paragraph 7.12.

7.9 Two handed process charts

The two-handed process chart is a process chart in which the activities of the operators' hands or limbs are recorded in their relationship to one another.

As the name indicates, this chart, as illustrated in Fig. 7.9. concentrates on the simultaneous activities of the two hands of an operator working at a definite work place (involving no or a little body movement) as in assembly or bench work. This is especially useful where a quick study is required to simplify the method by effectively using both the hands. There can be two types of two handed process charts.

a. Those indicating only operation, inspection and storage like the outline process charts.

b. Those indicating all the activities including movements and delays also like the flow process charts.

7.10 SIMO charts

It may be noted that the two-handed process charts are usually not charted against a time scales. And if they are drawn to a time scale in finer details generally indicating the therblig symbols, they become simultaneous motion charts or SIMO charts (Fig. 7.10). An additional thin column is provided for each hand, indicating the therblig either by its abbreviation or by its shading or by its color. Devised by Gilbreth, it graphically represents in the form of

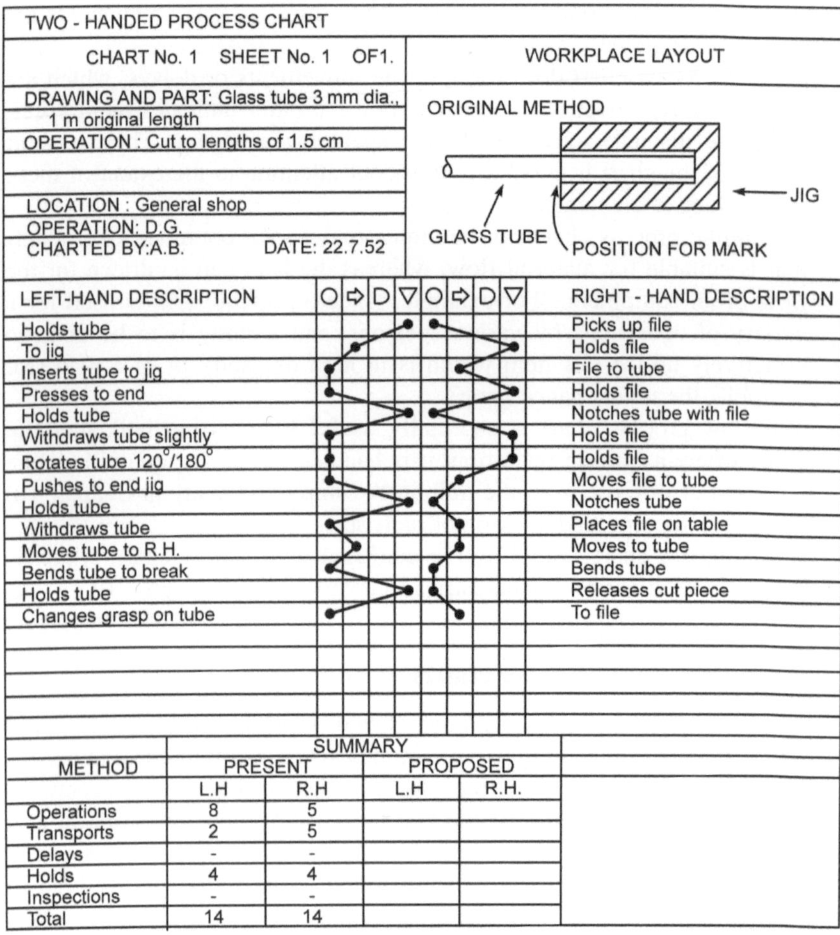

FIGURE 7.9 Two handed process chart. *Courtesy: ILOs Introduction to Work study.*

therbligs, movements of each limb of the operator under study. A more elaborate definition can be given to SIMO charts as under.

> *SIMO chart is the detailed systematic presentation of the method of work performed by the body members of the worker, usually working at a fixed workplace, as recorded by the motion picture.*

Uses of SIMO charts

1. Inefficient motion pattern of the operatives and the ineffective times can be identified and any violation of the principle of motion economy can be easily noticed.
2. Balanced two-handed motions with coordinated foot and eye motions can be achieved.

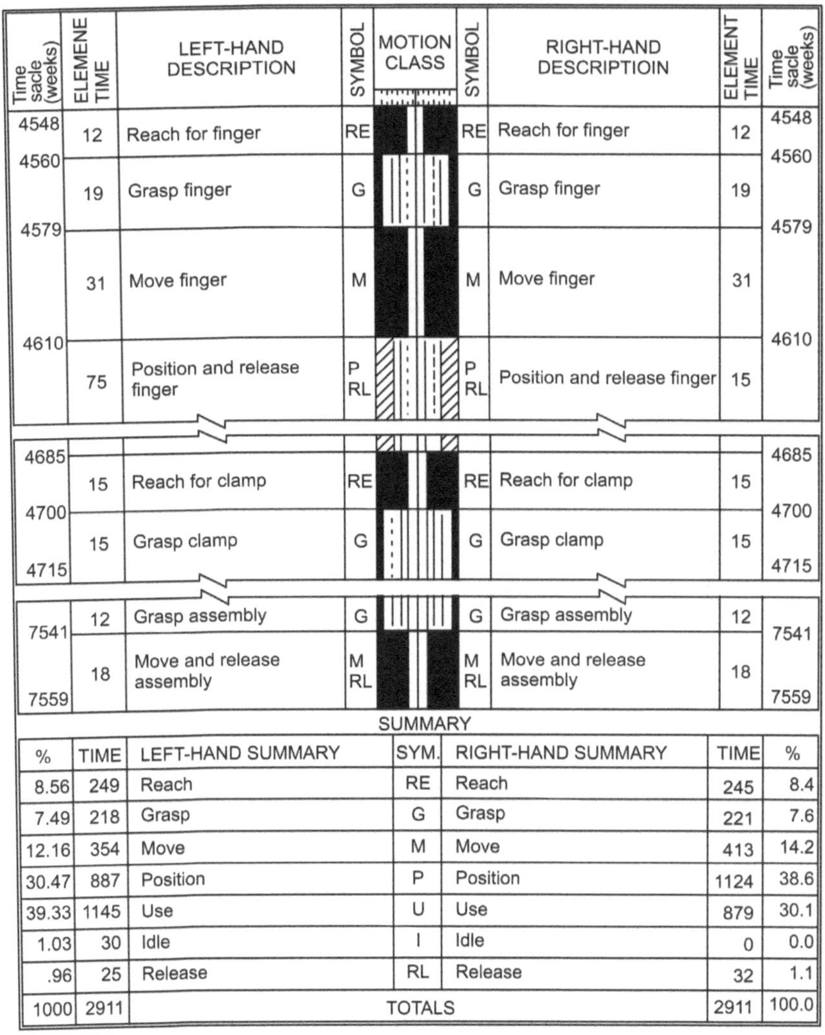

SUMMARY

%	TIME	LEFT-HAND SUMMARY	SYM.	RIGHT-HAND SUMMARY	TIME	%
8.56	249	Reach	RE	Reach	245	8.4
7.49	218	Grasp	G	Grasp	221	7.6
12.16	354	Move	M	Move	413	14.2
30.47	887	Position	P	Position	1124	38.6
39.33	1145	Use	U	Use	879	30.1
1.03	30	Idle	I	Idle	0	0.0
.96	25	Release	RL	Release	32	1.1
1000	2911		TOTALS		2911	100.0

FIGURE 7.10 SIMO chart. *Adapted from: Niebel BW, et al.* An introduction to methods, time study and wage payment.

3. The chart thus helps in improving the method of doing an operation resulting in a smoother, more rhythmic work cycle that keeps both delays and operator fatigue to the minimum extent.

7.11 Multiple activity chart

A multiple activity chart is a chart on which the activities of more than one subject (worker machine or item of equipment) are recorded on a common time scale to show their interrelationships, as illustrated in Fig. 7.11.

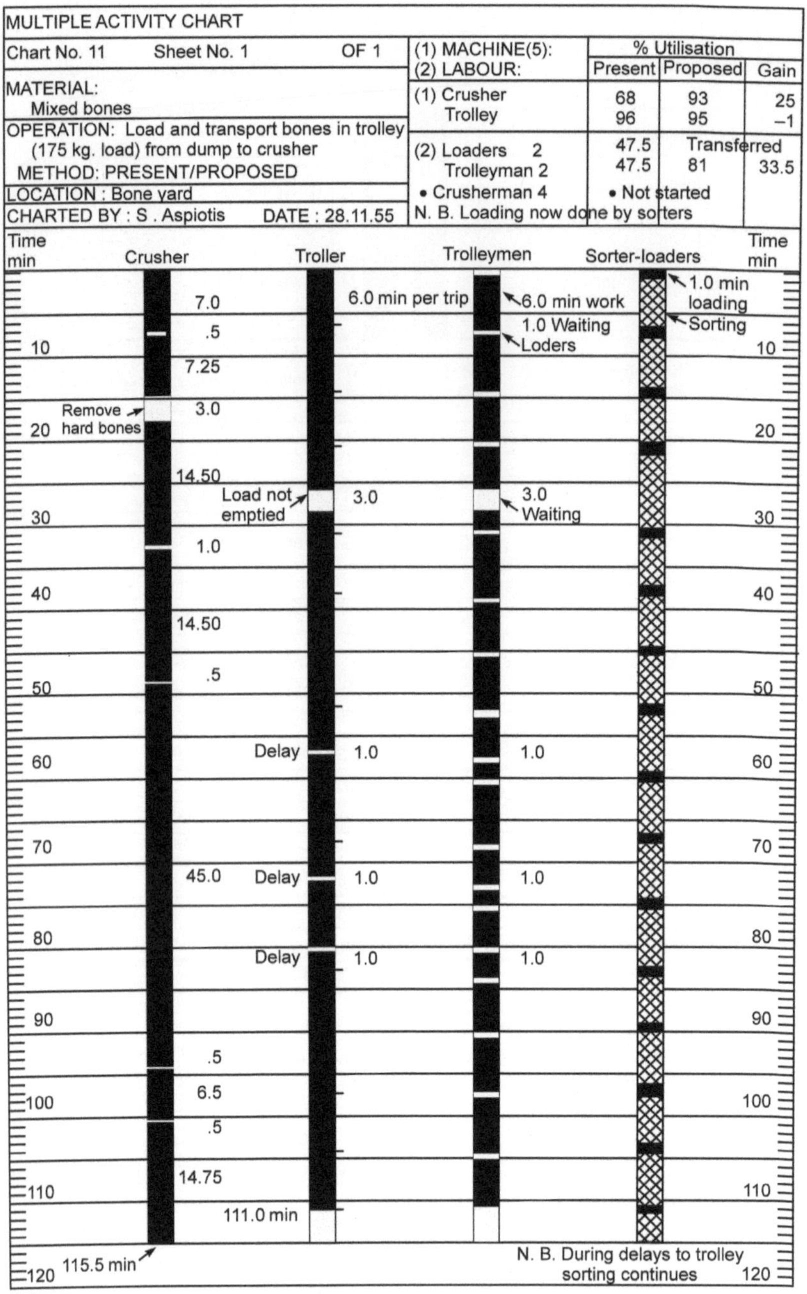

FIGURE 7.11 Multiple activity chart (man-multi equipment type). *Courtesy ILOs Introduction to work study.*

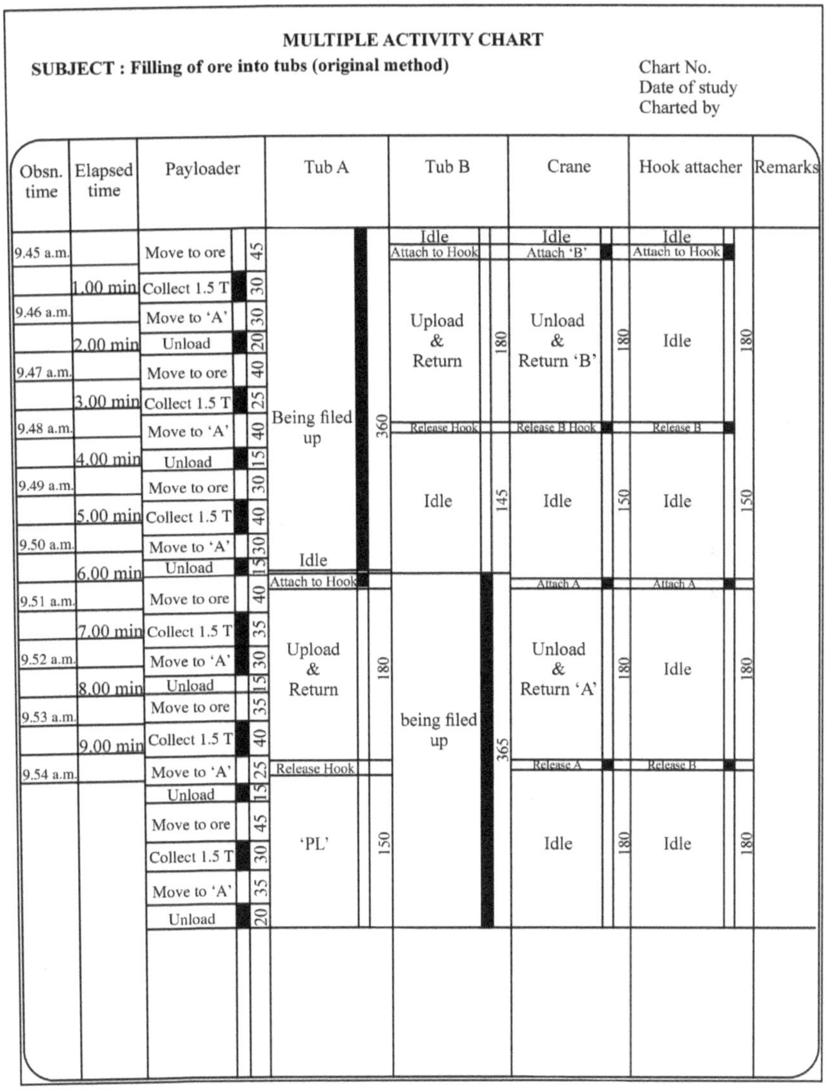

FIGURE 7.12 Multiple activity chart for the crane operation at a seaport.

Another multiple activity chart per Fig. 7.12 represents crane operating in a seaport. The iron ore, which gets unloaded from the railway wagons as per our case study in Chapter 10, would later be filled into 2 tubs, each of 4.5 T capacity, by payloaders. When one tub is being carried by the crane on to the ship, the other tub gets filled by the payloader. The Fig. 7.12 represents the activities of the payloader, the crane, the hook attacher and the 2 tubs.

These charts are especially helpful in group operations where you cannot time the job of a worker independently since his job is linked with that of the others. They can be drawn for any of the following.

- Man - machine (example: operations on an auto lathe)
- Man - multi machine (example: a weaver operating many looms in a textile weaving department)
- Multiman charts: (example: brick laying, maintenance gangs, etc.)
- Multiman - multi machine charts (example: gang operations involving many machines or equipment like loading of iron ore into ships).

The activities are represented by shading a vertical bar and the duration of the activity is represented by the length of this shaded bar against a time scale. The shadings are as follows:

- Operation time - fully shaded
- Idle time - blank
- Transporting time- crossed

Another illustration of the multiple activity is the working of a harbor gang operating in a ship hatch as a part of the operation of unloading ammonium sulfate from the ship. This material, which is imported in bulk is directly filled into the ship hatches in bulk at the originating port and due to its long sea voyage, gets hardened and needs to be chipped into small pieces to facilitate unloading by crane. Though this is not charted here, a study of the procrss helps us plan better methods to reduce this forced idleness. Reference can be made to the Chapter 10 for a detailed illustration of how this chart can be used or are handiling and other manual port operations to examine and develop optimal method.

7.12 Flow chart

A, B & C: Primary, secondary and cooling water circuits, **D:** Core reactor, **E:** Control Rods, **F:** Pilot Operated Relief Valve, **G:** Primary booster pump (EIW), **H:** Steam generator (boiler), I: Turbine - Generator, J: Heat Exchanger, **K:** Demineralizer, **L:** Condensate tank, **M:** Secondary water booster pump, **N:** Cooling Tower

This chart is generally used in process industry to indicate the flow of the product during various stages of the process. This is a combination of outline process chart and the flow diagram, where each operation is represented by the appropriate shape of the equipment as illustrated in Fig. 7.13. This gives a visual picture of the equipment as well as the operation sequence. Sometimes the equipment is represented by functional blocks, as illustrated in Fig. 7.13. Even in computer programming, flow charts are used to represent the series of decisions and the corresponding action taken. A typical such chart is illustrated on Fig. 7.14.

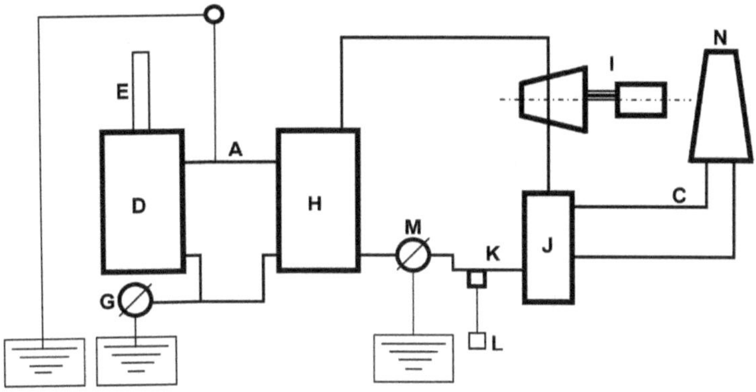

FIGURE 7.13 Flow chart of a nuclear Power plant.

Sl. no	Symbol	Explanation	Sl. no	Symbol	Explanation
1		Process/operation	16		Card
2		Alternative process	17		Punched tape
3		Decision	18		Summing junction
4		Data	19		Or
5		Predefined process	20		Collate
6		Internal storage	21		Sort
7		Document	22		Extract
8		Multi document	23		Merge
9		Terminator	24		Stored data
10		Preparation	25		Delay
11		Manual input	26		Sequential access
12		Manual operation	27		Magnetic disc
13		Connector	28		Direct access
14		Off-page connector	29		Display
15		Transfer of materials	30		Direction of flow

FIGURE 7.14 Symbols used in computer flow charts.

7.13 Computer process flowchart symbols

In computer software development, flowcharts are used to graphically represent the decisions and their results by a step-by-step diagram in boxes of various kinds.

These symbols, though rarely used in work-study, they are given here for comparison. However the reader is advised to distinguish between the terminology used in work-study and that in computer software development, especially the flow charts and flow process charts.

7.14 Flow diagram

A flow diagram is a plan, substantially to scale, of the factory or shop with the location of the machine, work place, etc. indicated. On this the movement of each product or component can be graphically represented.

As the definition implies, the flow diagram per Fig. 7.15 is a supplement to the flow process chart as illustrated in Fig. 7.6. This can also be used for the movement of men and tools. The general steps involved in drawing a flow diagram are

(a) Draw to scale the plan of the work area.
(b) Mark the relative positions of the machines and all equipment like benches, booths, racks etc.

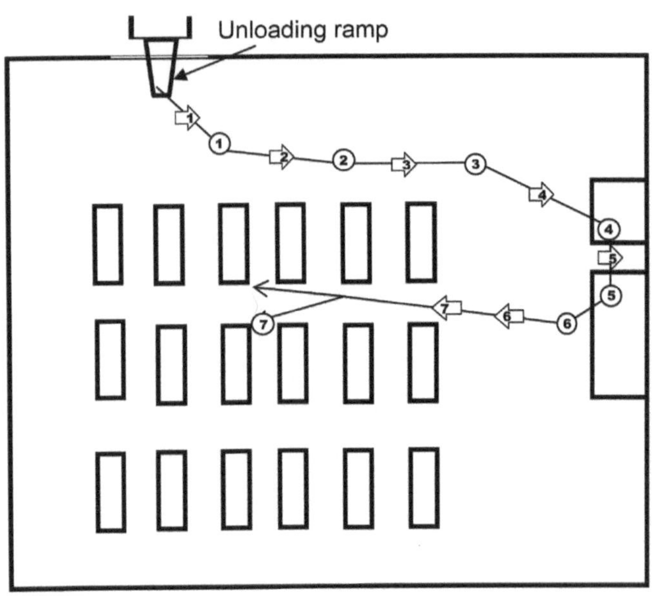

FIGURE 7.15 Flow diagram for a stores operation.

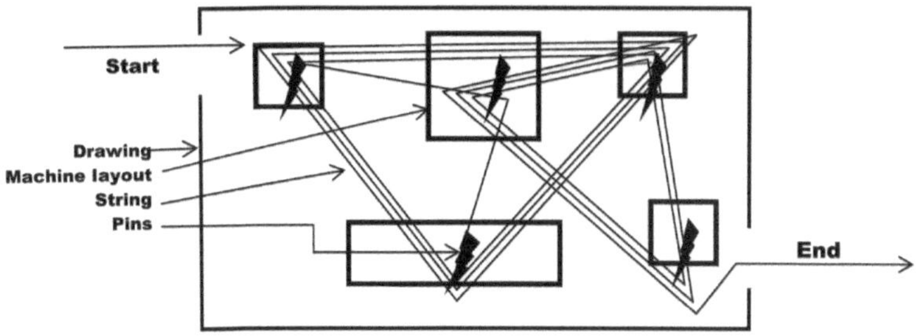

FIGURE 7.16 String diagram.

(c) From the different observations and recordings made, draw the path and direction of the movement of the material or men on the diagram. The paths of different components can be marked in different colors. Fig. 7.15 is an illustration for a typical flow diagram.

The difference between the flow diagram and the flow chart is that the former concentrates on the actual layout of the equipment and the flow of the materials, while the latter concentrates on the symbolic placement of equipment to suit the flow of the material. In other words, in the former, the flow suits the equipment layout while in the latter the symbolic equipment is shown to suit the flow.

7.15 String diagram

A string diagram is a scale diagram on which the movement within a given area is plotted by means of a continuous thread for the purpose of showing the frequency of movement between various parts and determining the total distance covered (Fig. 7.16).

The string diagram is in effect a flow diagram emphasizing the distance traveled by man or material. It identifies the places with greater concentration of movements. By using different colors of thread, the movement of more than one component can be represented in the same diagram.

The string diagram is thus used to investigate into the movement of two or more components or subassemblies that move between different workplaces in the same layout or when an operator attends to two or more machines or workplaces. The string concentration between departments also indicates the degree of closeness required between departments and suggests layout changes to reduce the total distance traveled. This diagram is helpful in plant layout.

7.16 Travel chart

The travel chart indicates the quantum of materials moved from one workstation or department to another in a table form. The quantum is

From department →

To department ↑

	A	B	C	D	E	F	G	H	Σ
A		6	8	15	20	16	4	7	
B	8		10	12	17	23	5	6	
C	13	6		-	5	8	3	15	
D	16	8	7		7	6	10	12	
E	-	4	9	10		8	12	14	
F	5	5	4	12	8		4	16	
G	-	20	25	4	12	10		4	
H	32	-	-	12	15	8	9		
Σ									

FIGURE 7.17 Travel chart.

measured in units like number of pallets or number of trips or tonnes moved or any other unit which well-defined and specified for the purpose of analysis. Fig. 7.17 indicates a typical travel chart involving eight departments.

7.17 Cyclograph

The cyclograph is a photographic representation of the path of the movement of the body members, especially of both the hands at the workplace. This is obtained by attaching a light bulb to both the wrists or the ring fingers of the operator and filming the movement by a camera whose shutter is kept open at very narrow aperture throughout the cycle. The developed film shows continuous lines of light indication the path of movement as illustrated in Fig. 7.18. This is similar to the landscape photographs of a marine drive or a busy road of a metro taken during the night, showing beautiful streaks of white and red lines on either side of the road, representing the head and tail lamps of the cars. This method was first used to study the movements of athletes used in 1890 by Marley and Gilbreth later developed it for work study.

7.18 Chronocyclograph

This is similar to cyclograph, but the light goes on and off at a frequency of 2–15 per second. The lights go off slower than they come on, that is the electric light bulbs are made to flash on quickly and die away slowly. This

FIGURE 7.18 Cyclograph.

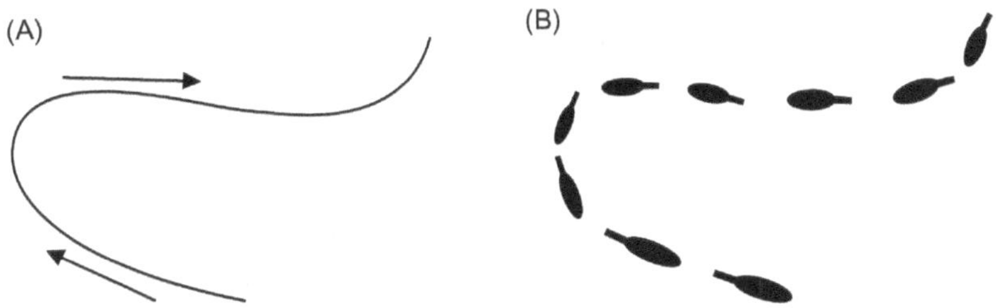

FIGURE 7.19 (A) Movement of the hand and (B) filmed spots of the Chronocyclograph.

produces a series of pear-shaped or comet like spots, on the film when developed. The tail of the spot indicates the direction of movement while the spacing between the spots indicate the speed of movement. A path trailed while studying a paper folding operation is given in Fig. 7.19.

Further Developments in Chronocyclography

- Filming with stereoscopic camera to get the 3D effect of the movement of the lights towards and away from the camera;
- Using color film with different colored bulbs for each of the two hands as well as other parts of the body to assist analysis

7.19 Memo motion photography

Here a time-lapse camera is placed over viewing the entire workplace and takes pictures at the rate of 1–6 per second without changing or advancing the film. This capture the different positions of the same part of the body in one photo, the distance between the same part being proportional to the speed. Sometimes the time lapse camera exposes the movement in subsequent frames so that an activity of 10–20 minutes can be compressed to

FIGURE 7.20 Screenshot of a video on the airport activity by time-lapse camera.

1 minute and the general pattern of movements can be surveyed rapidly frame by frame, from which the details of the wasted movements can be detected and steps taken to eliminate them. Besides saving analysis time, this also saves the film used up by 60%. This was developed 1946 by Marvin E. Mundel at Purdue University, as described in his book *Systematic Motion and Time Study* (1947).

7.20 Time-lapse camera video

While the above memo-motion photography is suitable for limited space in-house operations, large areas operations like the bulk material handling on the docks of seaports or airport operations, where several vehicles move near the airstrip, even while the planes are taking off, a video taken by a time-lapse camera would be useful to analyze the operations especially from the safety point of view. The time lapse camera exposes the movement in subsequent frames, either fast at 1–4 frames per second or intermittent at 4–8 s per exposure, so that an activity of 10–20 minutes can be compressed to 30–60 s and the general pattern of movements can be surveyed rapidly frame by frame. While this video exposes avoidable vehicular movements both from the cost and safety point of view, frame by frame viewing can aid further analysis. This analysis of the wasted and unsafe movements can be detected and steps taken to eliminate them. Fig. 7.20 illustrates a screenshot of a video taken by a time-lapse camera.

7.21 Conclusion

As highlighted earlier, recording, the second step of method study, is vital and provides the basis for the critical examination of the process. Several types of charts and graphical representation methods have been developed to aid analysis, each specific to its field of application. The latest to be developed is the time-lapse camera videos explained in paragraph 7.20. This author's grandson had taken several such videos out of photographic interest, but on viewing them, the author found them to be ideal for analyzing large area operations like airports, sea ports, etc., and made an attempt in this book, to develop it as a new recording technique naming it time-lapse camera video. Perhaps this is the first textbook on work study to represent this recording technique. If any reader is interested, such time-lapse videos showing the frenzy of activities at Boston and Philadelphia airports can be sent by e-mail.

Had such time lapse cameras been available during the sixties/seventies, the wagon unloading operation explained in the case study Chapter 10, could have been recorded with better clarity.

Criteria questions

1. Illustrate how symbols and charts explain a situation better than statements. (7.1)
2. Draw and explain the significance of each of the 5 process chart symbols. (7.4)
3. Enumerate the charts that do not use time scale. (7.5)
4. Distinguish between outline process chart and flow process chart. (7.7)
5. What is a SIMO chart? Illustrate. (7.10)
6. What is the difference between SIMO chart and multiple activity chart? (7.11)
7. Distinguish between Flow diagram and Flow chart. (7.14)
8. Distinguish between cyclograph and chronocyclograph. (7.18)
9. Distinguish between memo motion photograph and time lapse camera video. (7.19,7.20)
10. Explain and illustrate the significance of time lapse camera video. (7.20)

Further reading

1. ILO. *International labor organization - introduction to work study*. 3rd ed. 1977.
2. Kiran DR. *Production planning and control: a comprehensive approach*. BS Publications; 2017.
3. Maynard HB. editor. *Industrial engineering handbook*, 2nd ed. McGraw Hill Pub; 1963.
4. Carson, editor. *Production handbook*, Arnold Press; 1978.
5. Mundel ME. *Motion and time study, principles and practice*. 4th ed. Prentice Hall; 1988.
6. Shaw A. *The purpose and practice of motion study*. 2nd ed. Columbine Press; 1960.
7. Barnes RM. *Motion and time study*. 7th ed. John Wiley & Sons; 1980.
8. VID-20170630-WA0030.mp4 - Time lapse video on the activity at Philadelphia Airport tarmac. Video taken by myself at Philadelphia airport, 2017.

Chapter 8

Examine and develop

8.1 Significance of generating alternative solutions

Having selected the job and recorded all the data relating to the existing procedures, we have now to analyze this data and develop alternatives and viable solutions methodically and systematically to achieve simplification and overall cost reduction. In the earlier phases, especially during recording we might have stumbled upon alternatives, but it is in the development stage that our efforts to improve are concentrated. That means, at recording stage, generating of ideas is a by-product but in development stage, it is the objective. It is the free thinking and creativity in us that aid in the objective of generating the solutions. Hence it is imperative that we spend some time in understanding the significance of creative imagination and the principles of creativity.

Since the activity of developing alternatives happens simultaneous with the activity of examining, we discuss both together in this chapter.

8.2 Requirements for examining and developing

1. Accepting that the problem lies at the ineffectiveness of the present method of operation of the product or service.
2. Right attitude to solve the problem
3. Proper organization to coordinate the activity
4. Adequate knowledge and practice in using problem solving tools and techniques
5. Structured method of problem solving
6. Problem definition and analysis to be based on correct and complete data.
7. Solutions only to the cause and not for the symptoms of the problem
8. Implementing and continuous monitoring till consistent result is obtained
9. Overcoming resistance to change
10. Control system for reversible changes

Work Organization and Methods Engineering for Productivity.
DOI: https://doi.org/10.1016/B978-0-12-819956-5.00008-X

8.3 Significance of creativity in examining an operation

As explained in Chapter 1, during the 19th century, Frederick Taylor propounded the Scientific Management principles from which industrial engineering had developed into a science of studying a process and improving it on a continuous basis. It is hence imperative that in the following paragraphs, we will talk about the principles of creativity which is the fundamental requirement for our SREDDIM stage of Examine for improving the method of operation.

Let us recapitulate what management thinkers said about creativity.

- Shakespeare said, 'Creative imagination is the gift that makes the man the paragon of animals,'
- Einstein went further by saying 'Imagination is more important than knowledge'.
- Alex Osborne put the same thing in a different way, 'Knowledge may be power, but can be more powerful if creatively applied'.
- Robert T. Ross, in his contribution on Suggestions Schemes in Industrial Engineering Handbook, edited by Maynard, termed it as the Idea Power.

Ideas are obviously the key to the solutions for all kinds of problems, whether of production, materials handling, advertising, selling, human relations or more significantly, the housewife's planning the monthly household expenditure or managing the kitchen.

The human brain has many powers.

- Absorptive power - the ability to observe and apply attention
- Retentive power - the ability to memorize and recall
- Reasoning power - the ability to analyze and judge
- Creative power -the ability to visualize, foresee and generate ideas.

By first two we learn and by the latter two, we think. A computer may be able to perform the first three better than the human, but the creative power is the specialty of the human brain.

Have you ever wondered why the engineers, after their Master of Engineering get a Ph.D. (Doctor of Philosophy) degree and not Doctor of Engineering? This is true all over the world, but the concept is based on Hindu philosophy -

'Esa sarveshu bhuteshu gudho atma na prakasate,

Drsyate tvagryaya buddhya sukshmathi suksmam darshibi'

Atma is Present in all beings, but hidden and therefore is not manifest. It can be realized, however, by the concentrated reasoning of those who have trained themselves in perceiving subtle and more subtle truths. This is the basic principle behind all the research analyzes and it is this philosophy that is highlighted in awarding Ph.D. degrees. This is more so in the industrial engineering techniques, whether method study or reengineering or any analytical process. We call it creativity, the basic principle of critical examination.

Subtle truth can also be alluded to mother pearl found only in deep seas. Nothing will come to you on its own. You have to go very deep and investigate thoroughly to achieve the pearl. This is the truth.

8.4 Creative methodology

In our chapter on method study basics, we have listed the procedure as SREDDIM. The creative application methodology also forms almost the same steps, which we may rename as below:

1.	Select	Orientation	-	Locating and selecting the problem
2.	Record	Preparation	-	Data collection
3.	Examine	Analysis	-	Breaking down the problem into sub-problem
4.	Develop	Ideation	-	Thinking of all the possible ideas and alternatives
5.	Define	Synthesis	-	Selecting and defining the optimal alternatives
6.	Install	Evaluation	-	Testing and modifying the results and solutions
7.	Maintain	Follow up	-	Ensuring that the new method is sustained

We numbered them one, two, three.... etc., but the sequence, as far as creativity is concerned, is not rigid. We may analyze them even during the evaluation. We may still be digging for facts even after ideation. We may start guessing even while preparing. This is further detailed in paragraph 8.1.

8.5 The principles of creativity

8.5.1 Divide and conquer

We are all aware how the East India Company has widened its roots in India by splitting the Indian rulers by their negative principle of Divide and Rule. Nevertheless, this principle can be viewed positively in creativity. After selection, split each and every problem into components and deal with each sub problem individually, especially for questioning procedure. This will broaden the scope for more alternatives. The idea is somewhat similar to elemental analysis.

8.5.2 Set quotas and deadlines for yourself

The practice of setting up a quota of a minimum number of ideas to be thought over before a definite time always helps in getting larger and larger number of ideas.

8.5.3 Let loose your mind

As illustrated in Fig. 8.1 the human brain has two hemispheres, the left and the right. Whatever the right brain imagines, the left brain applies logic and erases the illogical imaginations even before they are formed. That means the left brain prevents us from getting ideas. So we should forget the presence of the left brain, think of ideas about from the right brain even though they appear illogical and then apply logic with the help of the left brain.

It hence follows that we should let loose our mind. Do not put any constraints to your thinking. Even if an idea appears silly, doesn't matter, list it. First think up and then judge. When it comes to thinking, try to act as if you have two different personalities, first a blind thinker and then a judge, only one at a time. The point is, even the silliest idea if analyzed further and tailored down, may lead to an ideal solution.

This is also called the brainstorming, wherein you want to include the opinions of all members of the team. This is explained more in detail in Chapter 11 on Kaizen.

8.5.4 Blue sky thinking

A new idiom that has found place in English vocabulary is Blue Sky Thinking, which means to get creative ideas out of the sky that are not limited to current thinking or beliefs. This expression stems from opening one's mind as wide as the blue sky, or let loose your mind as the previous paragraph emphasizes. London's Guardian Newspaper lists the following 10 blue sky ideas that changed the course of history:

1. Plato's Philosophy,
2. Freud's Theory of Unconsciousness,
3. Theory of Universe by Copernicus and Galileo,
4. Newton's Universal Gravitation Theory,
5. Einstein's Theory of relativity,
6. Descartes's theory of 'I think, therefore I am'.

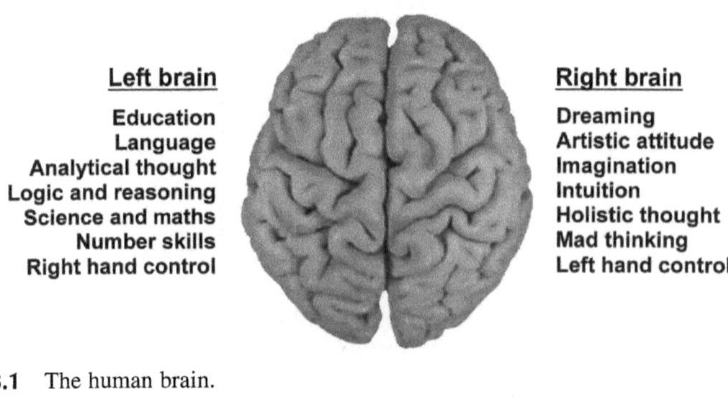

FIGURE 8.1 The human brain.

7. Marx's Analysis of capitalism,
8. Adam Smith's Laissez Faire Concept,
9. Women's Liberation, and
10. World Wide Web.

8.5.5 Two heads are better than one

This is the popular expression that has given birth to brainstorming which is a kind of group activity that is used to generate a lot of ideas. Participants are encouraged at the beginning to think of and list ideas, even if they sound silly or far-fetched as explained in the previous paragraph.

Here brainstormers are posed to a group of thinkers. Each and every idea put forward by them would be listed. It would be surprising to note that the number of ideas generated by a group would be more than the sum of all the ideas that each would be able to think of, if left alone. Paragraph 8.6 explains thus technique in full detail.

8.5.6 Question each and every detail

This is the cream of creativity. Rudyard Kipling said in his famous poem;

I had six stalwart serving men,

They taught me all I know,

Their names were What and Where and When

and Why and How and Who

This simile is similar to the Japanese metaphor, one husband and four wives where the four Wives (W) are **What, Where, Why** and **When** and the one Husband (H) is **How**.

Question each and every detail. Why? Why? Why? It would be emphasized without hesitation that a majority of the activities in our day to day life are based on this principle. 'Why'is the most significant and is the very basis for industrial engineers. It may also be stressed that while creativity is an art that depends upon the IQ, it can always be developed by the preceding guidelines and also by constant practice.

To illustrate the above, consider the simile that Satyabhama, the third wife of Krishna is the most powerful among his wives, and likewise, Kaikeyi, the third wife of Dasaratha is the most powerful among his wives. These 2 similes show that the third question viz 'why' is the most powerful one.

Industrial engineers, as a result of their years of experience in continuous improvement, never speak of the best method, but only of the best available method or the best method now desired. Even the Japanese word Kaizen speaks of continuous improvements as compared against innovation, which is a one-time improvement. In short, every time there is manual work involved, there is a continuing opportunity to improve the method.

If this concept is accepted, the method is examined and critical questions are put with an open mind, all the resistance to changes cited below and also referred to in Chapter 20, do not from the obstacles for the goal of methods improvement.

- "Not practicable for the operating staff"
- "Let us get back to realities"
- "Why change it? It is still working all right"
- "Our place is different"

The person who constantly asks questions and takes nothing for granted would certainly generate new ideas of doing the work if done in a systematic procedure. The questioning technique is the means by which the critical examination is conducted, each activity being subjected in turn to a systematic and progressive series of questions while the questioning attitude is a state of mind which takes nothing for granted during the investigation of a procedure or an operation. This systematic questioning results in determination of solutions based on facts and guards against the influence of emotions, opinions, habits or prejudices.

8.6 Brainstorming

Wikipedia defines brainstorming as a group or individual creativity technique by which efforts are made to find a conclusion for a specific problem by gathering a list of ideas spontaneously contributed by its members. In principle, you pose brainstorms to a group of thinkers. List out each and every idea put forward by them. It would be surprising to note that the number of ideas generated by a group would be more than the sum of all the ideas that each would be able to think of if left alone. Hence the expression two heads are better than one, as explained in paragraph 8.5.5, is the main principle behind brainstorming.

Osborn, who calls this as Ideative Efficacy, says that the two principles that contribute to this are (a) to defer judgment and (b) reach for quantity. This has given rise to the four factors of brainstorming which are,

1. Focus on quantity
2. Withhold criticism
3. Welcome unusual ideas
4. Combine and improve ideas

8.6.1 When to use brainstorming

1. Brainstorming helps your team generate a large number of ideas and to determine possible causes and or solutions to problems.

2. To aid planning out the project steps and to identify the different steps in implementing the project.
3. Brainstorming helps you on deciding which problem to work on and where many ideas need to be generated in a relatively short period of time.
4. When you want to include all opinions, and round Robin brainstorming as explained. below will helps to ensure equal participation in an idea generating session.

8.6.2 Freewheeling vs round robin

- Freewheeling, when anyone who has ideas is allowed to say them aloud. They are listed as they are said.
- Round Robin, where everyone takes a turn to offer ideas

8.6.3 Techniques of brainstorming

Wikipedia suggests the following methods of brainstorming

1. **Nominal group technique:** Participants are asked to write their ideas anonymously. Then the facilitator collects the ideas and the group votes on each idea.
2. **Individual brainstorming:** Here each person writes his ideas individually with his own word association and at the end all the papers are collected and studied.
3. **Group passing technique:** Participants sit in a circle, each writing down one idea, and then passing the piece of paper to the next person for adding one more ideas, till all participants write one idea each.
4. **Team idea mapping method:** Each participant is brainstormed individually, and at the end, all the ideas are charted together.
5. **Breaking the rules technique:** First the group decides on formal or informal rules that govern a particular process, and then try to develop alternative rules and methods.
6. **Directed brainstorming:** Each participant writes one idea in his paper and all such papers are randomly swapped among the participants. Till each paper is filled in by all the participants.
7. **Guided brainstorming:** Same as the breaking the rules method, but with the facilitator explaining some constraints in the mindset and also in time allowed.
8. **Question brainstorming:** First, participants are asked to develop questions by brainstorming, and then in a second round, write down the solutions and consolidate them.

8.7 Six thinking hats

Edward de Bono put forth a thinking process called Six Thinking Hats, to separate thinking into six clear functions and roles. Each thinking role is identified with a colored symbolic thinking hat. By mentally wearing and switching hats, you can easily focus or redirect thoughts.

1. The White Hat calls for information known or needed.
2. The Yellow Hat symbolizes brightness and optimism. Under this hat you explore the positives and probe for value and benefit.
3. The Black Hat applies judgment, and is called the devil's advocate or why something may not work. Spot the difficulties and dangers; where things might go wrong. It can be likened to the left half of the brain and probably the most powerful and useful of the Hats, but can become more problematic if overused.
4. The Red Hat signifies feelings, hunches and intuition. When using this hat you can express emotions and feelings and share fears, likes, dislikes, loves, and hates.
5. The Green Hat focuses on creativity, the possibilities, alternatives, and new ideas. It's an opportunity to express new concepts and new perceptions. This may be likened to the right half of the brain or 'Let loose your mind' as per paragraph 8.5.3.
6. The Blue Hat is used to manage the thinking process. It's the control mechanism that ensures the Six Thinking Hats guidelines are observed.

8.8 Other continuous improvement techniques

Continuous improvement can also be effected by other concepts like CREW, 3 M, DFSS and SMED which are explained further in detail in later chapters.

8.9 Primary and secondary questions

Primary questions cover the first stage of the questioning technique, when the very purpose, the place, the sequence, the person and the means of doing the operation are systematically questioned. Secondary questions cover the second stage of the questioning technique, during which the answers to the primary questions are subjected to further query to determine whether possible alternatives of place, sequence, persons and/or means are practicable and preferable as a means of improvement over the existing method. Alluring to paragraph 8.5.6, we can say that what, when, where, who and how are the primary questions, whereas why and what else are the secondary questions. This is where the mind is let loose and ideas created before we apply reasoning commenting upon the feasibility of each alternative. Fig. 8.2 illustrates a

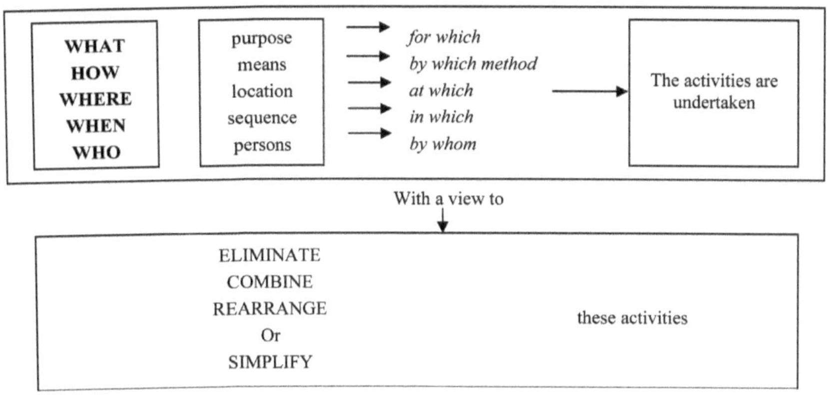

FIGURE 8.2 Primary Questions.

Name of the Organization : **Name of the Operation** : **Location** :	
WHAT Description of the operation in 2 or 3 key words	**WHY** Explain the need for this operation
HOW A. Indicate the equipment provided B. Explain the method adapted in detail. C. Indicate safety provisions.	**WHY THAT WAY** Explain the significance of each step
WHEN A. Indicate after which significant event. B. Indicate before which significant event C. Indicate the frequency.	**WHY THEN** Explain the reason for each indication
WHERE Indicate the exact location and area of the operation.	**WHY THERE** Explain the reason for each indication
WHO A. State the gang size B. Specify skill required C. Day/night working. D. Males only or male/female	**WHY THEY** Explain each indication

FIGURE 8.3 Critical examination chart. *Also Refer to the illustration in Fig. 10.2.*

Charted by : Supervised by : Date :		
WHAT ELSE List all alternatives, by letting loose your mind	**COMMENT** Comment on each alternative	**WHAT SHOULD** Choose the best alternative.
HOW ELSE List all alternatives, by letting loose your mind.	**COMMENT** Comment on each alternative	**HOW SHOULD** Choose the best alternative.
WHEN ELSE List all alternative sequences, by letting loose your mind.	**COMMENT** Comment on each alternative	**WHEN SHOULD** Choose the best alternative.
WHERE ELSE List all alternative locations, by letting loose your mind.	**COMMENT** Comment on each alternative	**WHERE SHOULD** Choose the best alternative.
WHO ELSE List all alternative locations, by letting loose your mind.	**COMMENT** Comment on each alternative	**WHO SHOULD** Choose the best alternative.

FIGURE 8.3 Continued.

systematically structured methodology for the critical examination, integrating the primary and secondary questions.

Fig. 8.3 (see after page 125) is the critical examination chart that should be prepared for each operation under study. Fig. 10.3 of the case study chapter illustrates how to fill up this chart.

8.10 Checklist for operation examination

Stegemerten and Geitgey, in their contribution on Operation Analysis in the third edition of Industrial Engineering Handbook, edited by H.B, Maynard, has suggested the following checklist while examining and analysis the method of operation.

1. **Purpose of Operation**
    - Is the operation necessary?
    - Does the operation accomplish the intended result?
    - Can the operation be eliminated by doing a better job on preceding operations?
    - Can the material supplier perform the operation more economically?
    - Can the operation accomplish additional results to simplify preceding operations?
2. **Design of Part**
    - Are all parts necessary?
    - Could standard parts be substituted?
    - Does design, permit least costly processing and assembly?
    - Will design allow eventual automatic?
3. **Process Analysis**
    - Can operation being analyzed be eliminated?
    - Or combined with another?
    - Or be performed during the period of another?
    - Is the sequence of operation the best possible?
    - Should the operation be done in another department to save cost or tooling?
4. **Inspection requirements**
    - Are tolerances, allowances, finish and other requirements necessary? Or too costly or suitable for the purpose?
    - Should statistical quality control be used?
    - Is the inspection procedure effective and efficient?
5. **Materials**
    - Consider alternative size, suitability, straightness and condition.
    - Can cheaper material be used?
    - Will tool modifications permit use lighter materials and thinner sections?
    - Would a more expensive material lower the machining and processing costs?
    - Is packaging suitable?
6. **Materials handling**
    - Can incoming materials be delivered directly to the workstation?
    - Can signals like lights and bells be used to notify the material movers that the material is ready to be moved?
    - Should cranes, gravity conveyors, tote pans or special trucks be used?
    - Consider the layout with reference to the distance moved.
    - Are containers correctly sized?
7. **Work Place layout, setup and tool equipment**
    - Is the arrangement of work area, location of tools materials optimal?
    - How are drawings and tools secured?
    - Can setup be improved?
    - Are the machine adjustments and trial runs are optimal executed?

8. **Methods**
    - Are hand motions symmetrical?
    - Are parts transferred between hands?
    - Is more detailed motion study needed?
    - Has safety is been considered?
    - What is the working posture?
    - Does the method follow the laws of motion economy?
    - Is the lowest class of movement used?
    - How does it compare with other operator on the same or identical job?
    - Can foot operated mechanism be used?
9. **Working conditions:**
    - Consider the following for improvement
    - Light
    - Heat
    - Ventilation
    - drinking fountains
    - Washrooms etc.

Apart from this checklist reference may be made to the appendix that details such checklist put forth by ILO in its book on the Introduction to Work study.

8.11 Develop

Though Develop is listed as a separate step of method study, it is inseparable with the step 'Examine' and both are generally done together, as explained in the opening paragraph of this chapter. This is where each of the generated alternative is considered with logic and weighed and evaluated for the economy, technical feasibility and other factors. A typical critical examination chart is illustrated in Fig. 8.2, wherein the first and second columns correspond to primary questions the third corresponds to the secondary questions, the fourth to the logic while the fifth column corresponds to Develop. This step determines the following as the last column of the format:

- What is to be done?
- How and by what means it should be done?
- Where should it be done?
- When should it be done? and
- Who should do it?

At this stage, we have initiated the development of alternative methods, may be one or two or three, and our proposals should be defended by a report that is to be submitted. After all, our suggestions have to be approved by someone else, may the top management or the production manager or even the supervisors, before we can install them.

The report should satisfy the following:

Give the details of the existing and proposed methods with full justification for the proposal,

(a) Precede by a summary of the recommendations and cost savings,
(b) Reflect the systematic procedure or compare the relative costs of material, labor, overheads etc. of the present and proposed methods, and the savings affected,
(c) Show the estimated cost of installation of the new or modified equipment, layout etc.
(d) Indicate the road map of action, that is 'who should do what and when' in the process of implementation and
(e) Remember to discuss the report with the persons concerned, especially the production manager, before the final submission.

8.12 Some quotations on change

- Change alone is eternal, perpetual, and everlasting.
  1. *Arthur Schopenhauer*
- Change begets change. Nothing propagates so fast.
  1. *Charles Dickens*
- If you want things to stay as they are, things will have to change.
  1. *Giuseppe di Lampedusa*
- Be the change you wish to see in the world.
  1. *Mahatma Gandhi*
- Every organization has to prepare for the abandonment of everything it does.
  1. *Peter Drucker*
- Without deviation, progress is not possible.
  1. *Frank Zappa*
- Great discoveries and improvements invariably involve the cooperation of many minds.
  1. *Alexander Graham Bell*

8.13 Conclusion

Right from Taylor's days, SREDDIM worked as the mantra of continuous improvement, enabling the global industrial production to attain high level of efficiency. Today KAIZEN, a Japanese term, but almost a synonymous concept of creativity is the same mantra to achieve the same goal of attaining high level of efficiency, as can be seen in Chapter 11.

Criteria questions

1. Discuss the significance of creativity in examining an operation. (8.3)
2. Relate Hindu philosophy of research to creativity. (8.3)
3. What is meant by SREDDIM? illustrate. (8.4)
4. What is the meaning of Divide and Conquer with respect to creativity? (8.5)
5. What is blue sky thinking? How is it related to the expression 'Let loose your (you) mind'? (8.5)
6. What is brainstorming? When it should be used? (8.6)
7. Distinguish between Freewheeling vs Round Robin. (8.6)
8. What are primary and secondary questions with respect to creative examination? (8.9)
9. Distinguish between the steps 'Examine' and 'Develop'. (8.11)

Further reading

1. Chang RY, Neidzwickin ME. *Their book continuous improvement tools*. Wheeler Publishers; 1998.
2. Krick EV. *Methods engineering*. John Wiley & Sons; 1962.
3. Imai M. *Kaizen: the key to Japan's competitive success*. New York: McGraw-Hill; 1986.
4. International Labour Organisation. *Introduction to work study*. 3rd ed. ILO; 1979.
5. Stegemerten, Geitgey. In: Maynard HB, editor. *Operation analysis: industrial engineering handbook*, 3rd ed. 1963.
6. Kart, Rosenzwig. *Organisation and management: a systems approach*. McGraw-Hill; 1970.
7. Spreigel, Lansburgh. *Industrial management*. John Wiley & Sons; 1966.
8. Barnes RM. *Motion & time study*. 7th ed. John Wiley & Sons; 1980.
9. Kiran DR. *Total quality management: an integrated approach*. BS Publications; 2016.
10. Kiran DR. *Production planning and control: a comprehensive approach*. BS Publications; 2017.
11. Kiran DR. *How to be more creative*. Journal of Rallis India.
12. Kiran DR. *Bulk materials handling at Madras port*. A PG Project; 1968.
13. Kiran DR. *Method study: a necessary tool for productivity improvement - a case study*. Industrial Engineering Journal; September 1982.
14. Kiran DR. *Material layout planning* . Industrial Engineering Journal; 1980.
15. Kiran DR. *Workstudy in transport sector, proceedings of work study seminar*. Tanzania; March 1982.
16. http://www.debonogroup.com.

Chapter 9

Method study - define, install and maintain

9.1 Define

In Chapter 7 we have studied the various recording charts that should be drawn for analysis and development. The same charts like outline process charts, flow process charts, process charts and flow diagrams, whichever were drawn for, the existing method should now be drawn and charted for the proposed operation procedure also. Each and every operation shall be detailed with reference to the procedure, tools, jigs, inspection gauges, etc. so that the shop personnel can perform the operation exactly in the way you set it without any wrong interpretation.

Apart from the above charts and graphic presentation, Standard operation procedures (SOP) as further detailed in paragraph 9.2 shall be written down. This also can be called Operative Instruction Sheet or Work Instructions. While procedures provide a general view of the higher-level steps, Work Instructions are significantly more detailed.

Whatever may the terminology, the International Quality Standard ISO 9001 essentially requires the determination of processes (documented as standard operating procedures) used in any manufacturing process that could affect the quality of the product.

9.2 Standard operating procedure

Once you define a process, you must write down the detailed procedure of operation in the form of an easy-to-understand, step by step list of instructions to the operator, as explained in the previous paragraph, to help them remember and carry out their day to day work and also refer to in case of doubts. It ensures that all related operatives and other personnel know exactly what and how to perform the operation. This we may call Defined Operating Procedure, in line with our SREDDIMs 5th step 'Define'.

9.2.1 Definitions on standard operating procedure

Several websites have given their explanatory definitions on Standard Operating Procedure

Work Organization and Methods Engineering for Productivity.
DOI: https://doi.org/10.1016/B978-0-12-819956-5.00009-1

SOP is "detailed, written instructions to achieve uniformity of the performance of a specific function".

The International Council for Harmonization (ICH).

A standard operating procedure, or SOP, is a set of step-by-step instructions compiled by an organization to help workers carry out complex routine operations.

Wikipedia.

SOPs are written steps to explain good manufacturing practices (GMP), plant safety routines, financial controls to secure assets, or IT security measures that employees are to follow. SOPs are step by step procedures on how to do something that is critical to quality, critical to safe operations, or critical to security.

www.bizmanualz.com.

A Standard Operating Procedure is a document which describes the regularly recurring operations relevant to the quality of the investigation. The purpose of a SOP is to carry out the operations correctly and always in the same manner. A SOP should be available at the place where the work is done.

FAO Document Repository.

The development and use of SOPs are an integral part of a successful quality system as it provides individuals with the information to perform a job properly, and facilitates consistency in the quality and integrity of a product or end-result.

US Environmental Protection Agency.

Standard operation procedures (SOPs) document the steps of key processes to help ensure consistent and quality output.

http://study.com/academy/lesson/standard-operating-procedure.

Standard operating procedures (SOPs) are the documented processes that a company has in place to ensure services and/or products are delivered consistently every time. It shows a buyer that the company is "process driven" rather than "founder driven."

https://www.divestopedia.com.

9.2.2 Objectives of standard operating procedure

The Main objectives of Standard Operating Procedure is to

(a) Record the improved method for future reference
(b) Explain the new method to the operatives, foremen and management without any difficulties in interpretation.
(c) Act as an aid for training.
(d) Form a basis for time studies and standards setting.
(e) Help in preparing material warrants and planning of inspection gages, special equipment etc.
(f) Reduce job hazards and ensure work safety.
(g) Help new employees to answer questions without having to interrupt supervisors to ask how an operation is performed.
(h) Achieve uniformity of performance, thereby achieving efficiency, quality output.
(i) SOPs need to remain current to be useful. Therefore, whenever procedures are changed, SOPs should be updated and re-approved.

9.2.3 Linking SOPs to quality

US Environmental Protection Agency states that the development and use of SOPs are an integral part of a successful quality system as it provides individuals with the information to perform a job properly, and facilitates consistency in the quality and integrity of a product or end-result.

9.2.4 Categories of SOP

Apart from the Written instructions for the manufacturing operations, for which we are concerned, we may cite here how FAO Document Repository categorizes the SOPs related to other industrial activities as below, denoting with a letter or combination of letters.

1. F-SOP for fundamental SOP, which give instructions how to make SOPs of the other categories
2. A-SOP or APP-SOP for apparatus SOP
3. M-SOP or METH-SOP for analytical method SOP, which describe a complete testing system or method of investigation.
4. P-SOP or PROJ-SOP for procedure to carry out a special investigation (project)
5. Q-SOPs for Quality Assurance.
6. I-SOPs for receiving inspection and registration of samples
7. PROT-SOP for a protocol describing a sequence of actions or operations
8. ORG-SOP for an organizational document

9. PERS-SOP for describing personnel matters
10. RF-SOP for registration form (e.g. chemicals, samples)
11. WS-SOP for worksheet (related to analytical procedures)
12. COMP-SOPs for documenting how to deal with complaints

9.2.5 Benefits of SOPs

- Workers who are following established procedures have more confidence that management supports their actions
- Customers are entitled to dependable product or service delivery based on established and accepted measures - not on any one worker's personal values, and not varying from one geographic area to another.

9.3 Install

Once the process is defined and well understood, the next step is to implement the new method or install it by putting into practice as per the written standard practice. This is the basic purpose of the whole method study assignment from the management as well as the work study engineer's point of view. As a quotation says, 'Making theoretical change is easy; making real change demands careful planning'. Any method study can be deemed as successful when the actual changes are made as per the defined instructions. By leading the implementation of the proposals, the work study engineer will have an opportunity to test and modify the details based on the feedback from the initial stages of implementation.

The work study engineer's job is also to train, guide and advise the operatives and the supervisors till such time the new method is adapted without any problems. Fig. 9.1 is a typical learning curve, which indicates that the worker often requires long practice to obtain the highest and consistent speed of working. They may need reassuring, retraining and supporting through the acquisition of new skills. Sometimes a parallel running of old and new systems is adapted to enable the operator to recognize the benefits of the improved procedure.

9.3.1 The 4 steps of installing a proposed method

We can summarize that the step installation consists of 5 stages.

1. Plan the installation process
   (a) Order new plants or materials (if any)
   (b) Phase out the changes in the production process
   (c) Decide on the extent of redeployment
   (d) Introduce new documentation procedures in line with SOP.
   (e) Set new quality standards and test procedures
   (f) Prepare a detailed timetable for effecting these changes

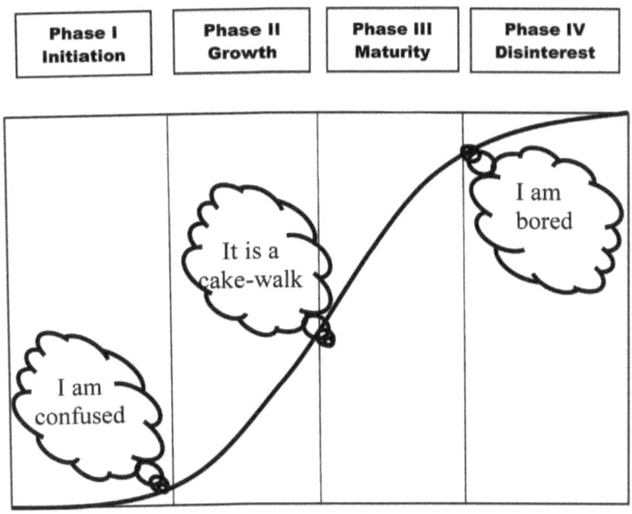

FIGURE 9.1 A typical learning curve.

2. Gain acceptance of the changes by the departmental personnel, right from the production manager to the supervisor.
3. Get approval of the top management.
4. Get acceptance by the concerned worker and or their representative, giving due attention to the fact of resistance to the change from them. Reference can be made to paragraph 21.6 of the chapter on Resistance to change.
5. Train and retrain the worker to operate in the new method. Even though the worker is experienced and normally gives over 100% output in the existing method, he cannot immediately attain the full standard performance in the very first trial of the new job.

9.4 Importance of training

In carrying out the first three stages stated in paragraph 9.2.1, the importance of preliminary education and training of all personnel concerned with the change, including the workers, supervisors and if necessary, the management in work study and its techniques, is significant. Their reception to the idea of change is more proactive when they know what and how the change is made than when they are merely presented with the result. Sometimes it may be necessary to decide if the savings by the new method justifies the training and retraining the old workers, who find it hard to come out of their old practices. In that case it is better to concentrate on training the new workers on the new methods, letting the old workers continue to work in the way they know.

9.4.1 Guidelines for training of the operatives

As emphasized in the previous paragraphs, it becomes imperative that the operatives are continuously trained on the new work methods. The following may be noted in this direction

1. Use films to demonstrate the old and the new methods. Films are particularly valuable when retraining.
2. Develop the habit of doing the job in the correct way. Train to follow a numbered sequence illustrated on a chart.
3. Understand and appreciate the learning curves in giving attention to the trainees as per their needs.
4. In the first stages of learning, rests between periods of practice should be longer than the periods of practice themselves.
5. When the operative has begun to grasp the new method and to pick up speed, rest periods can be very much shorter.

9.5 Maintain

The new method after successfully installed should not be allowed to slip back into the old method during the course of time or introduce elements not allowed for. It is absolutely wrong to presume that the work study man' job is finished the moment their proposal is approved and installed. There is a general tendency to drift away from the installed method. It is hence essential for the work study engineer to keep checking that the new method is being properly followed as per you plans, and that the desired results are being brought about.

He should maintain the proposed new method for fairly a long period till the new method gets set. It is imperative he keeps visiting the workplace and enquire for any difficulties in following the developed method and sincerely solve any of the difficulties they have been facing. The supervisor must be encouraged to understand the new method and to assist the workers in case of problems, but at the same time, the work study man must keep visiting to solve the difficulties. ILO has specifically recommended to maintain close contact with the job until the progress is satisfactory.

Sometimes the operative himself may offer certain suggestions after working on the new method for some time. Some of these which are helpful and should formally be incorporated A methods audit can be used to formally compare practice with the defined method and identify such irregularities.

9.6 Conclusion

As highlighted in this chapter preparing and popularizing the standard operating procedure is as important as the development of the new method of operation. Also significant are the 2 final steps, Install and Maintain, which should not be overlooked by the work study engineer.

Criteria questions

1. What is meant by Standard Operating Procedure? (9.2)
2. What are the Objectives of Standard Operating Procedure? (9.2)
3. Numerate the different steps in installing a proposed method. (9.3)
4. Illustrate how you will install a proposed method. (9.4)
5. Discuss the Guidelines for Training of the operatives. (9.4)
6. What is the significance of 'Maintain' the last step of SREDDIM? (9.5)

Further reading

1. International Labour Organisation. *Introduction to work study*. 3rd ed. ILO; 1979.
2. Kiran DR. *Total quality management: an integrated approach*. BS Publications; 2016.
3. Kiran DR. *Production planning and control: a comprehensive approach*. BS Publications; 2017.
4. Kiran DR. *How to be more creative*. Journal of Rallis India; 1969.
5. Kiran DR. *Bulk materials handling at Madras port*. A PG Project; 1968.
6. Kiran DR. *Method study: a necessary tool for productivity improvement - a case study*. Industrial Engineering Journal; September 1982.
7. Barnes RM. *Motion & time study*. 7th ed. John Wiley & Sons; 1980.
8. Mundel ME. *Motion and time study - improving productivity*. 6th ed. Prentice Hall of India; 1995.
9. https://en.wikipedia.org/wiki/Procedure.
10. https://www.wikihow.com/Write-a-Standard-Operating-Procedure.
11. http://www.fao.org/docrep/W7295E/w7295e04.htm.

Chapter 10

Methods study as a necessary tool for productivity improvement - a case study

10.1 Introduction

In the previous chapters, we had detailed the various aspects and procedures of method study, which is the systematic recording and critical examination of the existing and proposed ways of doing a work, as a means of developing and applying easier and more effective methods of operations, thereby reducing costs.

We had also seen in Chapter 6 that the basic procedure of method study is represented by SREDDIM, which means,

Select
Record
Examine
Develop
Define
Install and
Maintain

This chapter illustrates a typical case study of applications of the method study analysis conducted on handling operations at a seaport.

10.2 The case study

10.2.1 Brief details of the operations

This case study attempts to illustrate the procedure employed during a project conducted by the author on method study of the unloading operation on iron ore from the railway wagons of Madras (now Chennai) Port at Jawahar Docks, from where iron ore is exported. The rough plan of the ore berth is given in Fig. 10.1. This quay is about 500 m long and can accommodate three ships at a time if berthed longitudinally. It is worth mentioning here that the quay, a part of the Jawahar Docks, is normally reserved for export of

Work Organization and Methods Engineering for Productivity.
DOI: https://doi.org/10.1016/B978-0-12-819956-5.00010-8

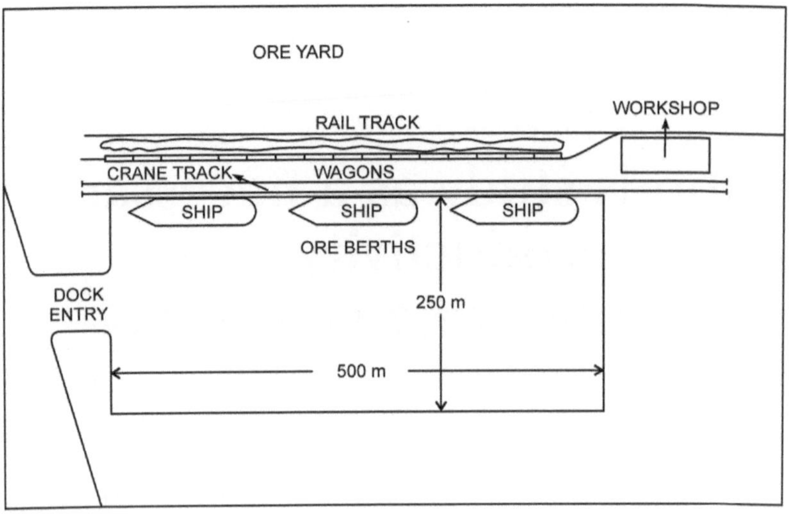

FIGURE 10.1 Layout plan of the ore berth (not to scale).

iron ore only, and in one calendar year 110 ships were berthed here together loading over 200,000 tons of iron ore.

The quay is provided with railway tracks about 20 m away from the wharf edge and the ore arrives in trains of 20 wagons, each of 22-ton capacity. The train is positioned on the quay, and a gang of 8 workers, plus a leader empty the wagon by shoveling off through the wagon door. Ten such gangs operate simultaneously emptying ten wagons at a time forming heaps of iron ore longitudinally along the railway track. The track is then cleared of the overflowing heaps of iron ore and the train moves by ten wagon lengths permitting the remaining ten wagons to be unloaded in a similar fashion. The track is again cleared by shoveling and the train moves off before the next operation of loading into the ship commences.

Now let us examine the procedure employed for conducting method study for this project.

10.2.2 Select

The selection of this operation for method study is based upon the following facts and considerations.

1. The project operations generally have long cycles with considerable amount of manual labor.
2. The value of the ore per tonne is very low compared to the cost of handling, and this fact is magnified by the large extent of manual labor involved. Hence, any improvement in the rate of unloading would considerably reduce the f.o.b (freight on board) cost of the ore.

3. The utilization of the costly machines and equipment, like the ships and the cranes is very low, as they cannot operate during the whole process of unloading from the wagons. Subsequently the actual loading rate is a small fraction of the rated capacity of the cranes.
4. The port management has since long been feeling the pinch of this uneconomical operation and are eager to consider any positive improvement on the system.

10.2.3 Record

The charts and diagrams indicated below fully incorporate the data required for the analysis of the present method.

1. Flow diagram (Fig. 10.1), which is the detailed plan of the work area, viz. the ore berth.
2. Equipment sketch (Fig. 10.2), gives a picture of the present method of the shoveling and the formation of the ore heap.
3. Flow process chart (material type) Fig. 7.6 of Chapter 7, indicating the 3 sub-operations performed. The remarks column indicates which operations have a large scope of improvement.

10.2.4 Examine & develop

The step EXAMINE forms the very backbone of method study. As explained in Chapter 8, each and every detail is questioned without bias or reservation as illustrated in the critical examination chart (Fig. 10.3), where each of the

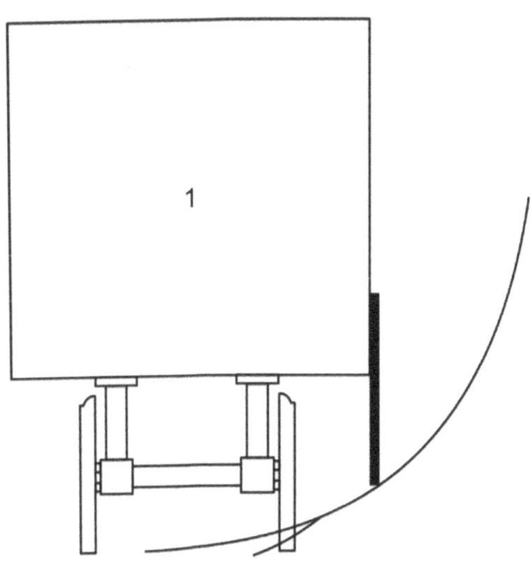

FIGURE 10.2 Door opening-existing.

Operation :: Unloading iron ore from wagons onto the berth Location :: Jawahar Docks of Madras Port (Now Chennai Port)	
WHAT Ore is being unloaded from the wagons	**WHY** A: The ore comes from mines in wagons B: The ore is be reloaded on to the ships
HOW A. Manually by shovels B. Method used: one gang of 8 men plus a leader work on each wagon, after opening the door to let some ore drop by gravity, all 8 climb the wagon and shovel off the ore. After emptying the wagon, the ore that falls on to the track, is cleared by shovel and by hand picking to enable the wagon to move away. C. No safety precautions are followed.	**WHY THAT WAY** A. No mechanical aid is provided. B1. It is the standard practice. B2. As the door is opened vertically down through 180°, (Fig. 10.2.1) the ore forms a heap covering the tracks fully, preventing wagon movement. C. No safety instructions provided
WHEN A. after the wagons arrive and position adjusted. B. Before the empty wagons move away C. 3 times a day.	**WHY THEN** A. Ore heap should be well located on the berth. B. Wagons cannot move unless the track is cleared. C. Unloading of one wagon and clearing the tracks take 30 min each.
WHERE A. At the work place towards the yard side of the wagons. B. All along the berth for 70 meters.	**WHY THERE** A. To prevent interference with the crane operation. B. All wagons cannot be unloaded into a single heap
WHO A. A gang of 8 men and a leader B. 8 unskilled and 1 skilled C. Day workers. D. All males	**WHY THEY** A. Gang size is fixed as a policy D. Unloading needs no skill. C. Work cannot be done at night D. Very laborious job.

FIGURE 10.3 Critical examination chart-case study.

five parameters, WHAT, HOW, WHEN, WHERE and WHO are discussed with regard to

1. The existing process
2. The expressed reason for doing so
3. The possible alternatives
4. The practical implication of each alternative
5. What should be the optimum process?

		Charted by :: D. R. Kiran Date 6th February 1968
WHAT ELSE a. Do not unload ore b. Unload only a part of the ore	**COMMENT** Objective not achieved	**WHAT SHOULD** Iron ore should be unloaded.
HOW ELSE A1. Throw out ore by baskets. A2. Provide chutes. A3. Use belt conveyers. A4. Install wagon tipping mechanism. B1. Keeping the door closed, shovel off ore over wagon. B2. Provide chutes sliding the ore 1 m away. B3. Open the door partly by 100° by special clamps (Fig. 10.2.4.) B4. Above plus additional fork-hook attachment to give an extension of 2 m (Fig. 10.2.5). B5. Above may be supported on legs and wheels. C. Give safety instructions on positioning of men during the operation.	**COMMENT** A1. Uneconomical A2. Low height A3,4. Present berth does not provide for these* B1. Inefficient. B2. Not possible with present wagon. B3. Good, but less projection. B4. Most ideal B5. Legs useful, but wheels not needed. C. Conduct Job Safety Analysis	**HOW SHOULD** A. Unload manually by shovels. B. Open doors partly by 100° by special clamps plus an additional attachment per the diagram, to make the ore fall 1.5 m away from the track (Fig. 10.2.3.). C. Issue Job safety instructions.
WHEN ELSE A1. Before wagons arrive. A2. While the wagons are in motion B. Before clearing the track. C. More frequent.	**COMMENT** A1. Impossible A2. Impractical. B. Impossible. C. With the existing method, only five rakes can be cleared per day.	**WHEN SHOULD** A&B. As at present. C. With New attachment 6 wagonsper gang, or 60 wagons in 3 rakes per day can be unloaded, with 10 gangs.
WHERE ELSE A. At marshalling yard B. Directly in to the ship C. At concentrated places	**COMMENT** A. No purpose served B. Impossible C. Inefficient	**WHERE SHOULD** At the current workplace on the berth
WHO ELSE A. More persons per gang B. All skilled workers	**COMMENT** A. No purpose served. B. Unnecessary	**WHO SHOULD** The present gang of 8 men and a leader

FIGURE 10.3 (Continued).

It can be seen that while parameters1 and 2 as above form the step EXAMINE, parameter 5 forms the step DEVELOP, while 3 and 4 form a overlapping of the two steps. That is the reason why both the basic steps are illustrated together in Chapter 8 as also in this paragraph.

10.2.5 Critical questioning

This critical questioning highlights the unproductive labor involved in clearing the railway track caused by the overflowing ore heap. Consequently the recommendation arrived at as per HOW ELSE columns are (Figs. 10.4–10.5).

(a) To open the wagon door by only 100° instead of by 180° by use of special clamps (Fig. 10.6) fitted to the wagon brackets prior to unloading

(b) To obtain an overall projection of 1.5 m by the use of an attachment (Figs. 10.7 and 10.3) so that the heap is formed at least 0.5 m away

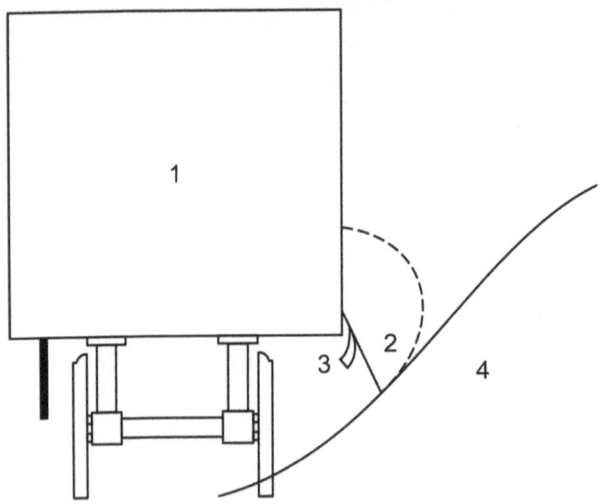

FIGURE 10.4 Door opening - with bracket.

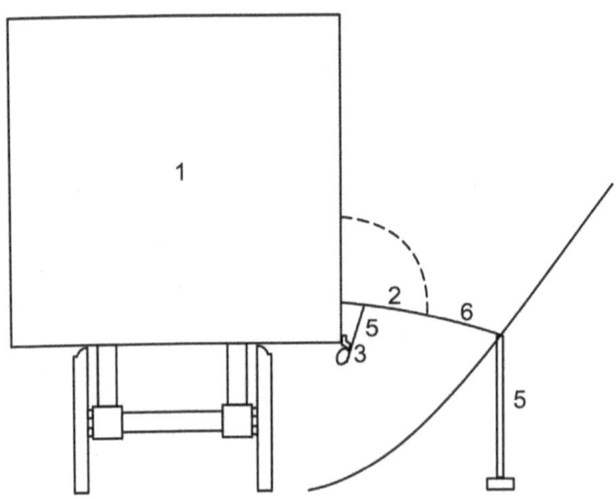

FIGURE 10.5 Door opening with attachment.

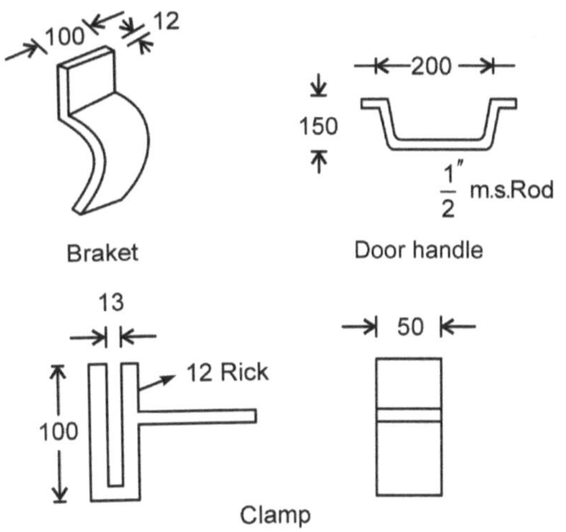

FIGURE 10.6 Bracket, handle and clamp.

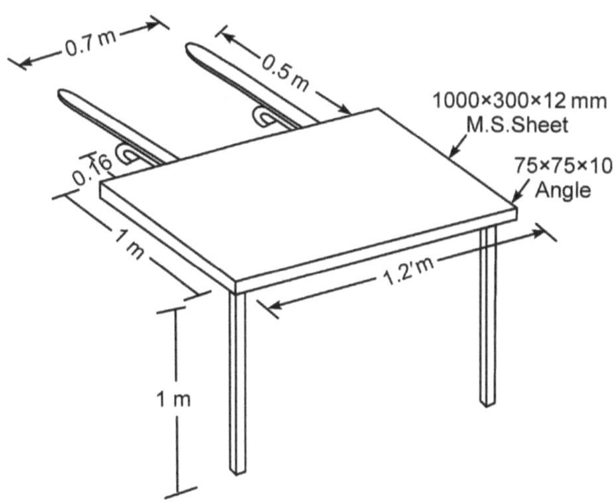

FIGURE 10.7 Special attachment.

from the track. The forks of the attachment (6) are inserted into the door handles (2) and the hooks (3) prevent it from slipping. Two legs (5) give the attachment additional support.

10.2.6 Capital investment for the attachment

Samples of the attachment and the clamps have been made at the port workshop at an estimated cost equivalent to Rs. 300.00 per attachment and Rs. 600.00 per pair of clamps, each with a service life of one year. Thus, for 10 wagons of 6 doors each, the annual capital investment is Rs. 36,000.00.

TABLE 10.1 Time saved per gang.

Parameter	Unit	Present	Proposed
Total unloading time	min	23.00	23.00
Track clearing time	min	15.00	03.00
Total cycle time before the wagon can move	min	38.00	26.00
Time saved per gang	min		12.00 (0.20 h)

10.2.7 Savings effected

The savings can be computed per the Table 10.1 below.

Thus, while the unproductive time has been slashed by 80% the total cycle time is reduced by 32%. This consequently has increased the effective rate of ship loading.

The total man hours saved per day is given by:

Hours saved per wagon gang $\times$ 10 persons per gang $\times$ 20 wagons per rake $\times$ 3 rakes per day $= 0.2 \times 10 \times 20 \times 3 = 120$ man hrs.

10.2.8 Define, install & maintain

Now that the alternative method is developed, the same has been presented to the port management in a detailed report form to seek their approval. Trials are taken with the fabricated samples by the port executives to confirm the exact time reduction in the track clearance. Their report is favorable, and the procedure is implemented during the subsequent month. The site has been visited during this month, when the proposed method has been found to be in practice without major problems.

The cost-benefit analysis, viz., the comparison of the rupees saved, with the cost of the equipment could not be made due to non-availability of the labor rates from the Port office. However, the management was quite happy at the total savings in terms of man hours and the increased ore unloading and loading rate and immediately agreed for its implementation. By increasing the net loading rate of the port handling system of the berth in question, this has resulted to the port management as well as the shipping lines in the form of reduced duration for the ship berthing time as detailed above.

It may be noted that subsequent to that iron ore unloading operation as described above, the ore would be filled by a pay loader into 2 tubs of 1.5 T capacity which are carried onto the ship by cranes. Though this too was a part of the 1968 project, as illustrated in the Fig. 7.10A of the chapter on 'Method Study-Record', it is not discussed in this case study.

It may be pointed out here that during 1960 the project objectives were limited to improving the unloading rate with the existing equipment. Nevertheless, this project initiated the Port's major project a decade later, of installing belt conveyors and tipping mechanisms for iron ore and other bulk materials handling at a new port extension at Ennore, around 6 km away. Yet another port extension has been developed for conveyorized handling of coal exclusively for the new North Chennai Thermal Power Station 15 km away.

10.3 Other case studies on application of creativity

(A) **Unbalanced workload on an assembly line:** A ceiling fan manufacturing company had a conveyor belt assembly line (Fig. 10.8) for its stator assembly, where the workloads among the different operatives on their line was unbalanced with one operative being occupied for 20% of the total cycle time, waiting for the work to come from the previous operator. During a study, the industrial engineer broke up all the operations into transferable elements of work and presented it before the operatives, who studied them and redistributed the elements among themselves so that each operative is equally occupied during the cycle time and the output of the line increased from 15 to 40 fan units per hour. This assembly line balancing can also be equated to an application of Heizunka, as listed in para 12.8 of Chapter 12 under the umbrella of Kaizen, though this author was not aware of this Japanese term in 1973 when the above project was implemented.

(B) **Fitness equipment at a peoples' park:** Look at the following photographs. The gym equipment used in most gymnastic clubs is so expensive that the patronage is poor due to the high cost of usage. However, a Chennai park purchased and installed simpler gym equipment that cost far less and can be used by the park users free of cost. Photograph 1 of Fig. 10.9 below, shows the costly gym equipment at a gym club with poor patronage compared to the park equipment shown in photograph 2. Photograph 3 shows a simple turning wheel that gives good exercise to the shoulders and back muscles.

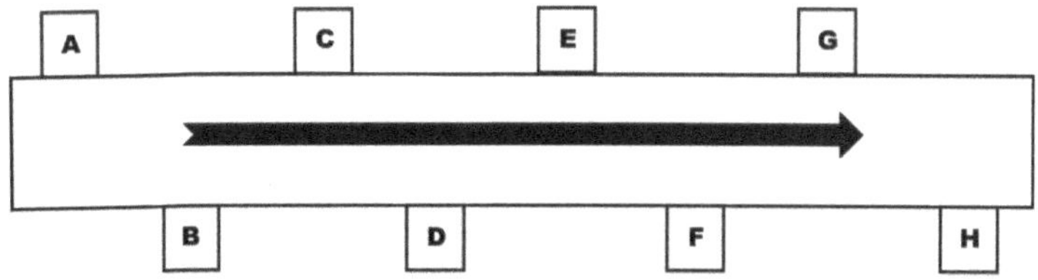

FIGURE 10.8 Illustration of assembly line balancing by method study.

Photograph 1: Gym at a spa

Photograph 2: Gym at the park

Photograph 3: The turning wheel for shoulder exercise

FIGURE 10.9 Critical examination technique applied to fitness equipment.

(C) **Problem of unfilled soap boxes in assembly line:** A soap manufacturing factory received complaints of unfilled soap boxes being passed on to the market. The factory management considered several alternatives to identify the unfilled cartons on their way out of the soap filing machine by suggesting sophisticated and expensive equipment like X-ray scanners with high resolution monitors. However, a shop level operative suggested mounting an air circulator pointing it to the conveyor belt, so that the empty boxes would fly off.

(D) **Automobile rear view mirrors:** Continual improvement can also be affected by minor unconventional innovative thinking. There was no concept of the rear-view mirrors when the first cars were introduced. One lady in the passenger seat was making up her face through her vanity mirror when her husband was driving the car. Suddenly she saw in the mirror something at the back of the car and drew his attention. Then she developed the practice of constantly looking into the mirror and informing him. Thus, the need arose, for a permanently fixed rear view mirror which today is the most used part of the car.

10.4 Conclusion

It can be concluded that there is no limitation to the coverage of fields for the application of methods study. Whether it is an engineering workshop or a chemical industry, a harbor or a hospital, a general office or a supermarket, an airport or other transportation center, or even in the spaceships, method study can effectively be used to bring about efficiency improvement, simplification and/or cost reduction. On the lighter side, we can cite the famous quotation of a Telugu language poet who said, 'no subject is unsuitable for poetry' (Kadedi Kavithaku Anarham). Likewise, we can say no field or activity is unsuitable for method study and continuous improvement.

Criteria questions

1. Give a brief report in 2 or 3 pages, an illustration of a simple method study project, indicating all the steps of SREDDIM.
2. Give a brief one paragraph note on any creativity application case study you know of, similar to, but different from, those of paragraph 10.3.

Further reading

1. I.L.O. *Introduction to work study*. 3rd ed. I.L.O.; 1979.
2. B.S.S. 3138. *A glossary of terms used in work study*. Bristish Standards Institute; 1992.
3. Summary of recommendations of the seminar cited above.
4. Kiran DR. *How to be more creative?* House Journal of Rallis Group; January 1975, p. 6–7.
5. Kiran DR. *Total quality management: key concepts and case studies*. Francis and Taylor; 2016.
6. Kiran DR. *Production planning and control, a comprehensive approach*. BS Publications; 2017.

Chapter 11

Kaizen and continuous improvement

11.1 What is kaizen's role in productivity improvement?

In Japanese,

	Kai	means	**Change**
and	**Zen**	means	**Good**
Thus	**Kaizen**	means	**Change for the good.**

Kaizen emphasizes continuous improvement as compared against innovation, which is a one-time improvement. We can say that Kaizen is the 20th century Japanese adaptation of industrial engineering, where creativity is the fundamental requirement. The Fig. 11.1 below indicates how kaizen attains high performance levels at no or marginal costs, as against innovations that need heavy investments. It implies that whatever we do can be improved continuously whether it is at the work place or at home. In other words, unlike the Management by Objectives (MBO) kaizen is a way of thinking and organizing everything-from the way you work and the way your team works together.

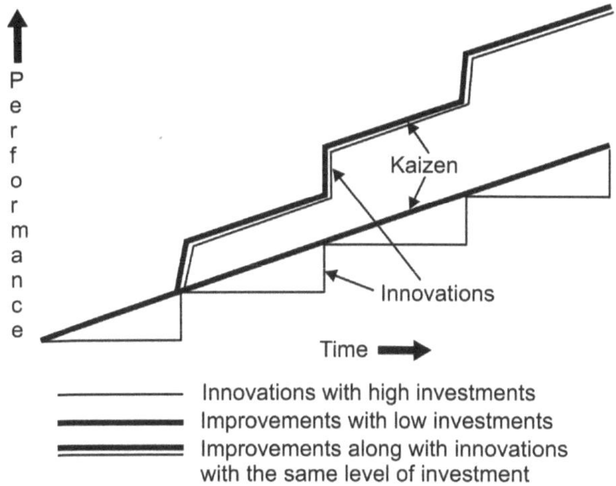

FIGURE 11.1 Kaizen vs innovation.

Work Organization and Methods Engineering for Productivity.
DOI: https://doi.org/10.1016/B978-0-12-819956-5.00011-X

11.2 Kaizen and creativity

In his book, Kaizen and the Art of Creative Thinking, Dr. Shingo, who originated Kaizen, as a Scientific Thinking Mechanism or basically as a habit-building approach, explains the following principles that reflect the ethos of Toyota's production system,

- Create an innovative company and culture
- Manage a creative environment
- Focus your resources for successful growth
- Let loose your mind for idea generation
- Appreciate and apply the true value of improvement, and
- Make full use of human potential

Kaizen is generally associate with the Toyota Production System, where it is used as one of its core business principles. We can cite here, Taguchi's famous narration that Toyota is so committed to continuous improvement that any worker on a Toyota assembly line could stop the line at any time to attend to a problem in production, correct an error, or suggest to management a better way to do things that reduces waste or improves efficiency. In fact, when the management is convinced that this stoppage has really led to higher productivity and higher quality, the worker gets rewarded instead of getting punished for stopping production. Perhaps this initiated the 'suggestion scheme' of the rest of the word.

11.3 Kaizen vs innovation

The western countries concentrate on innovation for development, Japanese maintain that it is Kaizen that pays more in the long run. They say that the effect of innovation wears off in a short time, but Kaizen supplements its impact and multiplies the returns. Masaki Imai, in his book *Kaizen: The key to Japan's competitive success*, differentiates them by two major management components: maintenance and improvement.

> *Maintenance requires activities as maintaining current operations of the company as technological, managerial, and operational standards. On the other hand, improvement can be broken down between Kaizen and innovation. It is logical because both concepts will lead to different improvements.*

He also observed that

> *The starting point for improvement is to recognize the need. This comes from recognition of a problem. If no problem is recognized, there is no recognition of the need for improvement. Complacency is the archenemy of KAIZEN.*

Changes have taken place all along the history, mostly for better, especially the post-World War era. The latter half of 20th century had seen tremendous changes not only in manufacturing processes and procedures, but also in management thinking.

Here are some of the differences between these two concepts:

1. Kaizen is by small improvements while innovation is by dramatic improvements.
2. Kaizen is in small steps while innovation is in big steps.
3. Kaizen requires only simple observations to discover simple problems resulting in a solution as an improvement effort. Innovation, on the other hand requires projects that may last several months to several years.
4. Kaizen is by continuous improvements while innovation is by periodic improvements.
5. Kaizen is in incremental improvements while innovation is in non-incremental improvements. For example, Kaizen attempts to reduce delivery time to 4 from 5 days and later reduce it to 3 days. Innovation focuses on process preplanning to ensure the delivery time to the same day.
6. Kaizen needs little investments vs innovation's large investments. This is clear because no sophisticated tools or equipment are needed for Kaizen, as stated in point no. 3 above.
7. Kaizen aims at improvements while innovation aims at rebuilding.
8. Maintaining Kaizen needs less effort while maintaining the innovation's needs great efforts.

11.4 Why continuous improvement?

Richard Y. Chang & Mathew E Neidzwick in their book *Continuous Improvement Tools* observed that

> *The single most destructive force in the move to improving the quality of American organizations today is the lack of commitment and understanding of how to make quality happen on the job.*

Thus, continuous Improvement is the hallmark of Productivity. As stated in Chapter 3 on Productivity, it has been present ever since man started living as a social animal and constantly strived to live better and provide better living conditions to his dependents. Einstein, Edison et al are known for their inventions and innovations, but even they had failed on their first attempt and finally succeeded because of their quest for continuous improvement and betterment.

11.5 Significance of kaizen in continuous improvement

- Kaizen is similar to methods improvement studies undertaken by the industrial engineers.

- It is a continuous improvement tool, and not a one-time approach.
- Its motto is 'Do 100 things 1% better, than doing one thing 100% better'
- It does not involve substantial investment.
- It aims at excellence at shop floor.
- It involves everyone in the factory.
- It involves identification of wastes involved in the production processes and operations.

11.6 How does kaizen improve productivity?

Kaizen the Japanese concept of creativity is simple and effective and can be applied to any task, goal or initiative, because

- Smaller tasks are more manageable, easier to plan for and allocate time for.
- Each task you complete is an achievement which is encouragement to take you further in achieving your ultimate goals.
- This method will improve productivity by eliminating wasted movements and operations.
- Using continuous improvement, you will improve the process.
- It is simple to use, simple to plan for, easy to follow and complete with the allocated time.

11.7 Juran's methodology

Juran observed in his famous Quality Trilogy, that among the three basic elements of quality management viz, Quality planning. Quality control and quality improvement, enough attention is being given for quality control, but not much attention for quality planning and quality improvement. He emphasized the importance of continuous improvement to Japanese industries, and this had ultimately resulted in a Japanese economic upsurge (Table 11.1).

11.8 Illustrations of kaizen application

Some illustrations of the continuous process improvements that can be undertaken are

1. By improving efficiency
2. By combining ideas from other fields
3. By trying to get things first time right
4. By eliminating rejections and rework by using new materials and technologies
5. By benchmarking with competitive process or products
6. By adopting JIT (Just in Time) systems of inventory
7. By constant customer feedback and review

TABLE 11.1 Juran's methodology for continuous improvement.

Step number	Juran's methodology	What it means
1	Proof of the need	Selection of a problem
2	Organization of diagnosis	Task force or quality circle
3	Diagnosis	Finding out the root cause
4	Breakthrough in the knowledge	Finding out solutions
5	Remedial action in the findings	Implementing solutions
6	Breakthrough in cultural resistance	Getting acceptance
7	Control at new level	Monitoring
8	Holding on to the gains	Defining the standard procedure

8. By using Quality Improvement Teams to solve problem
9. By using Six Sigma concepts
10. By adopting SPC and TPM techniques by using Quality Circles
11. By introducing Five S
12. By conducting CREW (cost reduction by elimination of waste) studies
13. By maintaining people to do their best by foolproof designing using Poka-Yoke
14. By training all employees as per their needs.

Reference may also be made to paragraph 10.3, which provide illustrations for the successful application of continual improvement which is the main emphasis of Kaizen.

11.9 Umbrella of kaizen

From the Japanese concept, Kaizen forms the umbrella covering several continuous improvement techniques, as illustrated in Fig. 11.2 below, in the same way as from the traditional concept, Industrial Engineering forms the umbrella covering several productivity improvement techniques, as illustrated in Fig. 4.9 of Chapter 4. This substantiates the similarity between Figs. 4.11 and 11.2.

11.10 Industrial engineering principles vs kaizen principles

In fact, it was during the 19th century that Frederick Taylor propounded the Scientific Management principles from which industrial engineering had

FIGURE 11.2 The umbrella of kaizen.

developed into a science of studying a process and improving it on a continuous basis. Thus, we can say that Kaizen is the Japanese adaptation of industrial engineering, where creativity is the fundamental requirement. Chapter 8 on Examine and Develop, discusses these principles in detail for better comprehension.

11.11 Conclusion

As said in paragraph 11.9, from the Japanese concept, Kaizen forms the umbrella covering several continuous improvement techniques in the same way as from the traditional concept, Industrial Engineering forms the umbrella covering several productivity improvement techniques. Right from Taylor's days, SREDDIM worked as the manthra of continuous improvement, enabling the global industrial production to attain high levels of efficiency. Today KAIZEN, the same manthra in different words, has enabled the global industrial production to attain a high level of efficiency.

Criteria questions

1. What is Kaizen? Relate it to creativity. (11.1)
2. Discuss Improvement vs innovation. (11.3)
3. Elaborate Juran's methodology on quality improvement. (11.7)
4. Explain and discuss the Umbrella of Kaizen. (11.9)

Further reading

1. Richard Y, Chang, Mathew E. *Neidzwick in their book. Continuous improvement tools.* Wheeler Publishers; 1998.
2. Imai M. *Kaizen:* the key to Japan's competitive success. New York: McGraw-Hill; 1986.
3. Kiran DR. *Total quality management: an integrated approach.* BS Publications; 2016.
4. Kiran DR. *Production planning and control: a comprehensive approach.* BS Publications; 2018.
5. Kiran DR. *How to be more creative - House Journal of Ralli Group.* January 1975.
6. Kiran DR. *Method study: a necessary tool for productivity improvement - a case study.* Industrial Engineering Journal, September 1982.
7. Kiran DR. *Material layout planning.* Industrial Engineering Journal, December 1980.
8. Kiran DR. *Work study in transport sector, proceedings of work study seminar.* Tanzania, March 1982.
9. Imai M. *Gemba kaizen.* 2nd ed. McGraw Hill Eduction; 2012.
10. Liker J. *The toyota way.* McGraw Hill; 2004.
11. Miller J, et al. *Creating a kaizen culture.* 1st ed. McGraw Hill; 2013.

Chapter 12

Terminology used in Japanese management practices

12.1 Introduction

During the post-industrial revolution era, the management thought and practices were developed mostly in the western countries, especially in USA. That is the reason why during the fifties and sixties we studied only the terminologies used by writers like Taylor and Peter Drucker.

However, during the post-World War II era, Japan emerged, as explained in Chapter 12, a strong industrial nation creating an awe among the western world in view of its highly successful management practices. This resulted in most of the management consultants and authors to understand and use the Japanese management terms.

It may be noted that most of these practices were in use in the western world also, but the emphasis in Japan is the importance given to the core worker, which was absent in the Western World. It may hence be said that the use of Japanese terms in place of the English terms created interest and indirectly helped young managers to understand and appreciate these practices better.

This chapter highlights the meaning and origin of some of the Japanese management terms used so as to provide a lucid insight into Japanese concept of World Class Management Practices.

12.2 Some of the terminologies cited in this chapter

- Kaizen
- Quality Circles
- Genchi Genbutsu
- Nemawashi
- Heizunka
- 3 Mu (Muda, Muri, Mura)
- 4 Wives and 1 Husband
- CREW
- 5 Management Objectives of Factory Management
- 5 Zu's

Work Organization and Methods Engineering for Productivity.
DOI: https://doi.org/10.1016/B978-0-12-819956-5.00012-1

- Poka Yoke
- Hanedashi
- Andon
- Jidhoka
- Chaku Chaku
- 5 S
- Gemba Walk
- Warusa Kagen
- Single Minute Exchange of Die
- Just in time
- Kanban
- Hoshin Kanri
- Nichijo Kanri
- Kata
- Six sigma
- Total Productive Maintenance
- Pecha-Kucha
- Dakaranani
- Kanso, Shizen and Shibumi
- OkyaKusoma

Some of these terms are further explained below.

12.3 History of development of japanese management practices

- Before World War II, Japan was not a highly industrialized nation. Most of the electrical and electronic goods were imported from US and Europe. So we can say Japan was playing a second fiddle to USA in commerce and trade.
- Japan's decision to side with Hitler alienated them against US and Japan's raid on Pearl Harbor infuriated US resulting in dropping of atom bombs on Hiroshima and Nagasaki, one of the most inhuman acts ever committed in the history of mankind.
- Consequently, Japanese wanted to pay back the Americans in their own coin.
- They knew it can't be in a war and hence decided to beat the Americans in world trade, by producing more quality goods and capture the international market currently reigned by Americans. Japanese being highly patriotic by nature, this desire has percolated into the minds of every national specifically into all categories of personnel in Japanese industry.
- Higher productivity became the initial buzz word. Yet they realized higher productivity without quality products can take the nation nowhere, the subsequent buzz word was high Quality. Production Quality in all aspects

of manufacture was given high priority. That's how the Japanese industry got the momentum for quality oriented higher productivity in order to capture not only the domestic but the international market.

- Around 1950, JUSE (Japanese Union of Scientists and Engineers) team visited USA to study USA's industrial practices. During their visit, they invited Dr. Deming, and subsequently Dr. Juran, to visit Japan and train their engineers.

12.4 Kaizen

As discussed in detail in Chapter 11, kaizen is a philosophy, emphasizing continuous improvement. While the western countries concentrate on innovation for development, Japanese maintain that it is Kaizen that pays more in the long run. They say that the effect of innovation wears off in a short time, but Kaizen supplements its impact and multiplies the returns (Fig. 12.1).

12.5 Quality circles

- Deming, an American consultant with Toyota was credited for the development of most of the Japanese techniques including introducing of the quality circles in 1950.
- The basic principle behind Quality Circles is to involve the grass root workers in all management decisions for change, by regular meetings and discussions among workers.
- Originally for controlling defects at the shop floor, quality circles later widened to include methods improvements, maintenance problems etc.

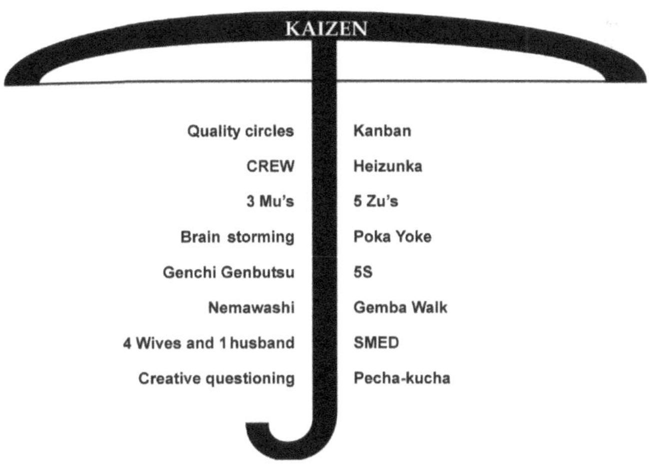

FIGURE 12.1 The umbrella of kaizen.

12.6 Genchi Genbutsu

Genchi Genbutsu literally means 'Go to the source to find the facts to make correct decisions'. In other words, it is the old industrial engineering principle to conduct a study and collect information on the job directly instead of resorting to only to available statistical data. This is somewhat similar to Gemba walk, where senior executives are involved.

12.7 Nemawashi

Nemawashi implies to make decisions slowly by consensus, thoroughly considering all options. This is the basic management principle cited as employee participation in Chapter 20 on Resistance to change

12.8 Heijunka

Heijunka implies to level out the work or to redistribute the work in such a way that every operator gets more or less the same workload. This is the very principle behind line balancing of assembly lines of small products like ceiling fans that are assembled by operators sitting in either side of the belt conveyor that moves the work in progress, as illustrated in paragraph 10.3. This line balancing involves

- Splitting each operation into transferable elements
- Timing of each element by work measurement. The term time study, which created a negative impression during the latter half of 20th century, as discussed in Chapter 2, is avoided, though in principle both mean the same.
- Considering the elements that can either be eliminated or redistributed to other operatives in the line so that each gets more or less the same workload in terms of the mean operational time.

12.9 3 Mu checklists

Kaizen practitioners have developed a system of checklists to help workers and management to be constantly mindful of areas of improvement, similar to the several checklists developed and used by industrial engineers all over the world since the fifties. In Japan, these are called the 3 Mu Checklists.

The 3 Mu's are

- Muda (Signifying waste)
- Muri (Signifying strain)
- Mura (Signifying discrepancy).

The fields where these 3 Mu's can be applied can be:

- Method of operation
- Process involved

- Facilities
- Jigs and tools
- Materials
- Production volume
- Inventory
- Place
- Manpower
- Technique
- Ways of thinking
- etc. . . .etc.

12.10 4M checklist

Propounded in the Western world, The 4M Checklist is similar to 3 Mu checklist but included here to provide a comparison between Japanese 3 Mu and western world's 4M.

(a) **Man (Operator)**

1. Does he follow standards?
2. Is his work efficiency acceptable?
3. Is he conscious of the problem?
4. Is he responsible? (Is he accountable?)
5. Is he qualified?
6. Is he experienced?
7. Is he assigned to the right job?
8. Is he willing to improve?
9. Does he maintain good human relations?
10. Is he healthy?

(b) **Machine (Facilities)**

1. Does it meet production requirements?
2. Does it meet process capabilities?
3. Is the oiling (greasing) adequate?
4. Is the inspection adequate?
5. Is operation stopped often?
6. Does it meet precision requirement?
7. Does it make any usual noises?
8. Is the layout adequate?
9. Are there enough machines/facilities?
10. Is everything in good working order?

(c) **Material**

1. Are there any mistakes in volume?
2. Are there any mistakes in grade?
3. Are there any mistakes in the brand name?
4. Are there impurities mixed in?

5. Is the inventory level adequate?
6. Is there any waste in materials?
7. Is the handling adequate?
8. Is the work in process abandoned?
9. Is the layout adequate?
10. Is the quality standard adequate?

(d) Method (Operation)

1. Are the work standards adequate?
2. Is the work standard upgrade?
3. Is it a safe method?
4. Is it a method that ensures a good product?
5. Is it an efficient method?
6. Is the sequence of work adequate?
7. Is the setup adequate?
8. Are the temperature and humidity adequate?
9. Are the lighting and ventilation adequate?
10. Is there adequate contact with the previous and next processes?

12.11 Four wives and one husband

This originates from a popular Japanese saying and explains the principle of questioning technique which is similar to Critical Analysis Technique or Cost Reduction through Elimination of Waste (CREW) adapted by industrial engineers world over even prior to the fifties.

- The 4 Ws (Wives) are **What, Where, Why and When**
- The 1 H (Husband) is **How**.

Among these, 'Why' is the most significant and is the very basis for industrial engineers. In fact, just to highlight the significance of why, this author prefers to choose the second letter H of Why to call it as the husband and the third letter W of How to group it under Wives.

Remember the famous poem of Rudyard Kipling?

I had six stalwart serving men,

They taught me all I know,

Their names were What and When,

And Where and Why and How and Who.

12.12 CREW

In contrast CREW (Cost Reduction through Elimination of Waste) is popular in the Western World. Though this is not a Japanese term, it is included here

TABLE 12.1 Waste categories under CREW.

Waste category	Nature of waste	Type of economization
Work-in-progress	Stocking items not immediately needed	Inventory management
Rejection	Producing defective products	TQM
Facilities	Having idle machineries	Increase capacity utilization ratio
Process time	High production costs	Method improvement studies
Production delays	Non-smooth flow of work in progress	Production planning & control
Down time of machinery	Excessive break down time and tool set up time	TPM, SMED, etc.
Expense	Over investing or over expenditure	Technical audit
Indirect labor	Excessive personnel	Effective job classification
Design	Products with more functions than needed	Value analysis
Operator talent	Highly skilled workers employed for routine operations	Effective job assignment

in view of its similarities with kaizen, etc. Canon has identified 9 waste categories indicated in the Table 12.1.

12.13 5 Management objectives of factory management

The five key points set forth by Mitsubishi Corporation, which are cited here in this category of Japanese management practices, are:

- Achieve maximum quality with maximum efficiency.
- Maintain minimum inventory.
- Eliminate hard work.
- Use tools and facilities to maximize quality and efficiency and minimize effort.
- Maintain a questioning technique and open-minded attitude for constant improvement based on teamwork and cooperation.

12.14 5 Zu's

In Japanese Zu, as a suffix means don't

like	*Math*	in Hindi,
or	*Vaddu*	in Telugu
or	*Vendam*	in Tamil.

The five don'ts or the things the operators should avoid doing with respect to defects are:

- Uketorozu meaning don't accept defects
- Tsurazu meaning don't make defects
- Baratsukasazu meaning don't create variations
- Kuriakalsazu meaning don't repeat mistakes
- Nagasazu meaning don't supply defects

12.15 Poka yoke

- Poka Yoke or AUTO-NO-MATION is based on the philosophy 'to err is human'. That is instead finding out who erred, find out why it happened and ensure it does not happen again.
- It ensures that the machine stops automatically whenever there is an error. That is the machine automatically says no to further operation, from which the term AUTO-NO-MATION is born. This is also called 'mistake proofing'. Examples are the CNC machines.
- It may be noted that even in 1950, the textile looms were provided with mechanical interlocking system that, whenever any thread snaps off during weaving, a thin reed supported by the thread slips down into the mechanism and the machine stops automatically.

12.16 Andon and hanedashi

Andon is an indication to stop work manually in case of any problem. While Poka Yoke involves automatic stoppage of the machine, Andon involves the manual stoppage by the vigilant worker.

Hanedashi is the use of auto ejection devices to unload the parts automatically after the operation is over. This is similar to poka yoke but is applied after the operation.

12.17 Jidhoka

Jidhoka is based on the philosophy that all individuals are responsible for the services they provide. Its purpose is to eliminate product defects by preventing, correcting, or drawing attention to human errors as they occur, It almost means same as Poka yoke. Both these terms were coined by Shigeo Shingo as a part of the Toyota Production system,

12.18 Chaku chaku

Chaku Chaku means 'Load, Load' referring to the positioning of all the machines as per the operation sequence and very close to each other. In other words it implies Product Layout.

12.19 5 S

5 S is a method for organizing workplace like shop floor or an office space. It advocates what to keep, where to keep and how to keep (maintaining, cleaning, etc.). It also instills a sense of ownership among the workers to be more accountable for their work place.

This forms a part of the Kaizen management, and emphasizes on simple tasks and work habits of the workers at grass root level, which when performed meticulously would result in substantial improvements and cost reduction. These simple tasks are often ignored by the similar workforce of the Western culture, whose attitude is that they have to do only specified jobs and these small tasks and responsibilities are of supervisors and managers. These 5-S's are:

- SEIRI - SORTING - Distinguish between necessary and unnecessary
- SEITON – SYSTEMIZING - A place for everything & everything in its place
- SEISO - SHINING - Keep the workplace clean
- SEIKETSU – STANDARDIZING - Maintain a good environment
- SHITSUKE – DISCIPLINE - Follow the rules of the company.

12.19.1 SEIRI (straighten up)

"Distinguish between the necessary and the unnecessary"

Few examples of Unwanted items are

1. Machines to be scrapped
2. Rejected materials
3. Expired goods.
4. Broken tools, pallets, bins and trolleys
5. Old notices
6. Scrap heaps near the machine.

12.19.2 SEITON (put things in order)

"A place for everything and everything in its place" (PEEP)

Few examples where SEITON should be applied are:

1. Unlabeled tool crip
2. Clustered shelves, lockers
3. Stores with no clear location system
4. Things on the floor.

12.19.3 SEISO (clean up)

"Keep the workplace clean"

Few examples of dirty work places that need cleaning are:

1. Dirty machines
2. Dust on products, parts and raw materials
3. Dirty jigs, fixtures, moulds and bobbins
4. Dusty walls, roofs, littered floor, etc.
5. Untidiness outside the factory.

12.19.4 SEIKETSU (personnel cleanliness)

"Maintain a good environment"

Few examples are:

1. Make it a habit to be clean and tidy, starting with your own person
2. Keep all visual control tools so that anyone can see and use them easily and conveniently.

12.19.5 SHITSUKE (discipline)

"Follow the rules of the company"

1. All employees must be trained in all aspects of 4-S. This must be done through classroom, actual practice and also by well-illustrated instructions and displays.
2. Supervisors must correct wrong practices every day on the spot.
3. If all members of an officer or workshop are not present when they are supposed to be, it is poor 5 S, punctuality is the backbone of 5 S.
4. Timeliness also applies to preparation of reports, filling out the charts, etc.
5. Wearing appropriate safety wear at all times is essential.
6. Name plates (badges) should be worn.

12.19.6 Shitsuke is the foundation for 5 S

As illustrated in Fig. 12.2 Shitsuke forms the heart and foundation for 5 S

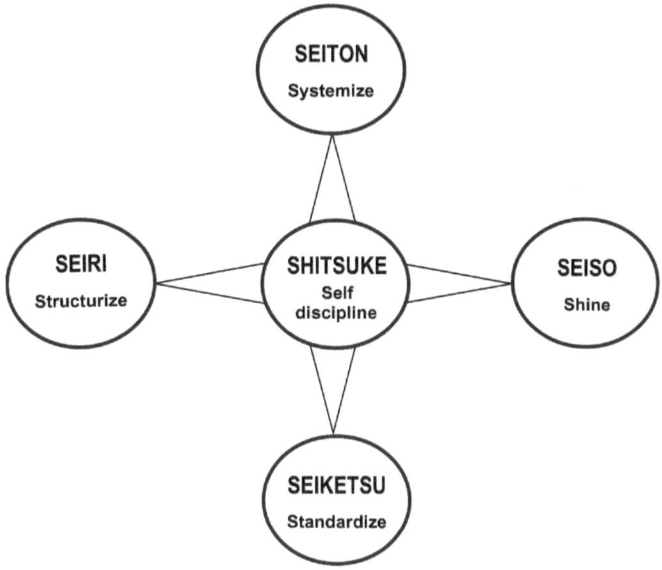

FIGURE 12.2 Shitsuke is the foundation for 5 S.

12.19.7 An easy way of remembering the 5 S terms

The appendix at the end of the chapter suggests you an easy way of remembering the 5 S terms and the order on which they appear.

12.20 Six sigma

Six Sigma is a business management strategy originally developed by Motorola, USA in 1981. However, since Japanese adapted this principle initially and were more successful than the rest of the world, and since the Japanese industries have been the benchmarks for Six Sigma concept, this is cited here to highlight its basic principle.

Since the 1920 the word 'sigma' has been used by mathematicians and engineers as a symbol for a unit of measurement in product quality variation. However, the engineers of Motorola in USA used 'Six Sigma' as an informal name for an in-house initiative for reducing defects in production processes.

The concept with which we use the term in statistics is different from that it is used in TQM, as can be clearly understood by the explanation given in Chapter 24 of the book *Total Quality Management, Key Concepts and Case Studies* by this author.

Hence it is important to call it as six sigma (at the most 6-sigma) and not by the symbol σ Also while referring to the extent of its application, to use the term 'level' or 'performance level' and not 'value'. That is to say,

'I achieved six sigma level quality' and not 'I achieved a quality of six sigma value'.

12.21 Gemba walk

Gemba walk is the practice of senior managers to tour the several places of the factory along with the concerned operatives with the basic purpose of identifying the areas of improvements. Each and every Muda noticed or suggestions offered by anyone for improvements would be recorded and analyzed and posted on the notice board to motivate the operatives. This is generally done during the afternoon hours or Sundays to create a friendly and holiday atmosphere may be followed by snacks or lunch. Preferably the Chief Executive Officer accompanies them to instill interest among the workers.

12.22 Warusa kagen

Warusa Kagen implies that things are not problematic now, but may soon develop into, unless controlled now. It is a caution noticed by a vigilant worker on the system and hence a starting point for several improvement activities.

This can be understood and remembered better by the Tamil metaphor - Warusaikku Aghum - (or Varusaku agunu in Telugu), both implying 'will be next in the line'.

12.23 Single minute exchange of die

Single Minute Exchange of Die (also known as SMED), is the Lean Manufacturing tool used to create very fast changeovers and setups to reduce machine downtime and increase throughput. SMED was developed by Shigeo Shingo and successfully introduced in Toyota Motors reducing machine changeover times from hours to less than ten minutes. The success of this program contributed directly to just-in-time manufacturing, which is part of the Toyota Production System. SMED makes Load balancing much more achievable by reducing economic lot size and thus stock levels.

12.24 Just in time

Just in time is a production and inventory control technique to ensure that the inventory level either as stocks in the store or as work-in-process on the shop floor is reduced to a minimum, almost to a zero level.

JIT purchasing is to ensure that the supplies are received in small quantities just in time for production, by establishing an agreement with vendors.

JIT on the shop floor is to ensure that each machine produces just when it is required for the next machine in quantities, not more than what is required. This is also called a pull-system of production. Kanban as explained below is a part of this concept.

Prior to the eighties, during the days of manual inventory control by Kardex system, the critical items with low stocks are identified by a thin red strip inserted in the card, so that procurement action can be initiated once a month to avoid stock outs. But today with the computerization such control can be done on a day to day basis even for most A and B items, enabling application of JIT concept to reduce the average inventory to 1 or 2 days level.

12.25 Kanban

While JIT is applied in production planning Kanban is applied in production control. But both are complementary and applied together to that extent Kanban has become synonymous with JIT system. We can say

JIT: KANBAN:: Production Planning: Production Control.

- Kanban, in Japanese language, means visible signboards, cards or chits.
- The Kanban can be a card, a container or an electronic signal.
- Every machine operator tends to produce only those quantities required for the next operation and keeps a Kanban in the container of the components as an indication to the next operator that the required semi processed material is ready. He slows down his pace for the next lot if the container is still not drawn by the next operator.
- When the subsequent operator finishes his operation, he draws the material returning this Kanban, forming a signal to produce a further lot of the required quantity.
- Hence this is also called the pull system of production.
- The underlying principle is that the needed parts should be received just in time for further processing.

12.26 Hoshin kanri

Hoshin Kanri is a Japanese term for strategic Planning.

Hoshin Kanri can be broken down into four parts,

Ho = direction

shin = shining needle, used in a compass

So Hoshin means progress towards a goal

Again Kan = control or channeling the progress

ri = reason or logic

So Hoshin Kanri means achieving the organization's direction, focus or goal, by logically controlling the progress.

In other words, Hoshin Kanri represents the management planning and control towards the achievement of the goal. It is a method devised to capture and cement strategic goals as well as to provide insight about the future and develop the means to bring these into reality. This is called strategic planning in the western world.

As Dr. Yoji Akao puts it,

With Hoshni Kanri, the daily crush of events and quarterly bottom line pressures do not take precedence over strategic plans, rather, these short term activities are determined and managed by the plans themselves.

12.27 Nichijo kanri

Nichijo means daily routine and kanri means management and control, similar to kanri of hoshin kanri. Thus nichijo kanri covers all the day to day aspects of operations planning and is complementary to hoshin kanri which refers to the long range or strategic planning.

12.28 Kata

Kata is a descriptive term for the organizational routines. It can be defined as behavior patterns, routines or habits of thinking and doing that are practiced over and over every day.

Kata as a term became popular in Toyota and provides a level of clear insight into the key behaviors underlying Toyota Culture in a way that can be easily understood and applied. Mike Rother in his book 'Kata in Toyota', refers to "improvement kata", "coaching kata", etc.

We may hence equate the term Kata (the story of the routines) to the Indian term Katha, but it is something more than a Katha. This is the very philosophy, which all the employees of Toyota breathe day in, and day out.

12.29 Total productive maintenance

Like any other Japanese practice, Total Productive Maintenance emphasizes that the base worker shall be entrusted with the task of performing the routine maintenance activities for the machine he operates. The maintenance problems too are discussed by the quality circles to ensure that the machine upkeep is more effective.

It may be noted that while the principles and procedures for preventive maintenance originated in the West, the Japanese emphasis of making the operator as the central focus for the routine maintenance combining with the 5 S practices made all the difference in the effective overall maintenance of the equipment. And this resulted in TPM.

Today world over, it has been realized that this Japanese practice of worker oriented routine machine maintenance is much more effective than the maintenance conducted by maintenance department workers.

12.30 Pecha-kucha

Literally meaning chatter or chit-chat, it emphasizes on the need to plan contents and time management of the Power Point or otherwise presentations thereby avoiding Pecha-kucha during the presentation. If you have 20 slides and only less than seven minutes to complete the presentation before discussions, the slides advance automatically before 20 s forcing you to limit presentation of each slide for 20 s only. This concept is hence called $20 \times 20 - 6.40$ (minutes). Started in Tokyo in February 2003 as an event for young designers to present their work, it has turned into a massive celebration and Pecha-kucha Nights are held in hundreds of cities around the world.

12.31 Dakara nani

Literally meaning 'so what?' it implies that while planning any slide, be prepared for the audience to question 'so what?' for every step and prepare yourself for a convincing response or edit the material accordingly.

12.32 Kanso, shizen and shibumi

These are the three vital elements of presenting your report in a meeting. Kanso means 'simplicity' that is achieving maximum with minimum means, using ideal concepts, visual elegance and perfect communication. Use of overelaborate designs and excessive refinement is avoided. Shizen means Naturalness in presentation of ideas that suit the particular audience. This point is also emphasized by Taichii Ohno in his book Toyota Production Systems. Shibumi means elegant simplicity in visual communication and graphic design.

12.33 Okya kusoma

This may not be a management practice as such, but means honorable guest, a term often used for the customer in Japan. Thus, this may be called a marketing strategy.

12.34 Conclusion

As explained in the synopsis, the author makes an attempt to enlighten on some of Japanese terminology and practices, the application of which has resulted in a major revolution in industrial management all over the world during the latter half of 20th century, next in importance only to the industrial revolution of the 17th century.

Of course during the early part of 21st century China has overtaken Japan in industrial success and maybe we have to retune ourselves to the Chinese Management terminology and practices. We should hope for the day when India overtakes the rest of the world in successful management practices and when every other country starts using Sanskrit words on management as their regular vocabulary on management terminology.

Criteria questions

1. What is the need for you to understand some common Japanese management terms? (12.1)
2. Trace the history behind Japan overtaking USA in commerce and trade. (12.3)
3. What is the significance of Kaizen in industrial engineering function? (12.5)
4. Explain the relationship between Heijunka and assembly line balancing with an illustration. (12.8)
5. What do you mean by the axiom $20 \times 20 - 6.40$? (12.30)
6. How do you distinguish between 6 and six sigma? (12.20)
7. What do you understand by JIT: KANBAN:: Production Planning: Production Control? (12.25).
8. What is the difference between Patrol inspection and Total Productive Maintenance? (12.29)
9. What advices the Japanese terms give you on power point presentation? (12.30)

Further reading

1. Kiran DR. *Total quality management, an integrated approach*. BSP; 2017.
2. Kiran DR. *Total quality management, key concepts and case studies*. USA: Elsevier; 2018.
3. Kiran DR. *Production planning and control - a comprehensive approach*. BS Publishers; 2018.
4. Kiran DR. *Japanese management practices*. A PP presentation; 2010.
5. en.wikipedia.org/wiki/Japanese_management_culture.
6. Weihrich H. www.usfca.edu/fac-staff/weihrichh/docs/management_practices.pdf.
7. Martin JR, et al. maaw.info/ArticleSummaries/ArtSumMartin92.htm.

Appendix

An easy way of remembering the 5 S terms

Look back at the list of 5 S - so many sei's! Seiri, Seiton, Seiso - or is it Seiso before Seiton?-Ooph, it is hard to remember these Japanese terms especially the order in which they appear.

Now let us see if the following illustration gives us is an easy way of remembering them!

A rolling Operation (abbreviated as Ri) is being done on a rolling machine by an operator called Krishna. After completing the operation he has to approach the switch box to switch off the machine. Let us call this switch box as 'K'.

That is, after completing the operation called 'Ri', Krishna has to go towards the switch box 'K'.

In other words,

The operation 'Ri' is done. So Krishna goes to Switch 'K'

Let us put these words this way -

Ri -- done -- so -- Krishna -- (goes to) Switch K

In Tamil script, 'T' and 'D'; are same and generally the nickname for Krishna is Kitchu. Again 'goes to' is not important here, but is written within brackets only to complete the sentence.

Hence we can rewrite this as,

Ri -Ton, - so- Kitchu- Switch K

By slightly modifying the spelling, this becomes

ri - ton- so - ketsu - Shitsuke.

Now add Sei to the first four

Seiri, Seiton, Seiso, Seiketsu and Shitsuke.

Summary:

Table - Summary of easy way of remembering 5 S terms.

Word as above'	Modified word	5 S word
Ri	Ri	Seiri
Done	Ton	Seiton
So	So	Seiso
Kitchu	Ketsu	Seiketsu
(goes to) Switch K1	Shituke	Shitsuke

You can practice reciting these words fast as you would do in the tongue twister games. Initially keep remembering the above illustration to check your order, and gradually you would master the terms and the order. One of the professor during the author's college days used to say,

'Even if you are awoken in the middle of the night, you should be reciting these words before you open your eyes.'

Chapter 13

Principles of motion economy

13.1 Introduction

In the previous chapters, we have seen how method study helps us to identify the unwanted movements around the workplace, thereby improving productivity. These unwanted movements we identified cover mostly the transportation of goods, tools, etc. as well as the overall body movements. However, if you closely observe the operative working in sitting position, you will realize that there are several finer movements of the individual body parts like hands, fingers, etc. which are performed unwantedly and unknowingly. These could be eliminated to achieve total reduction in the work content.

13.2 Basic body movements per Frank Gilbreth

Frank Gilbreth was the first to study these movements or motions during 1880 while observing a brick layer's operations. Later he refined these principles and applied them to find quicker and more efficient means of assembling and disassembling small arms in US Army.

Gilbreth named these as motion economy principles which form a set of rules and suggestions to improve the manual work in manufacturing and help to achieve productivity and reduce physical exertion of an operative at the workstation at micro or sub-micro level and eliminate wasted movements of the body parts, ease operator tasks, reduce fatigue and minimize work stress. Paragraph 13.5 explains and illustrates these therbligs further.

13.3 Categories of motion economy principles

During the 1930 Ralph M. Barnes codified these principles which are still valid and gave some analytical explanation to these Gilbreth's principles in his book, *Principles of Motion Economy - Some More Details.*

He broadly classified the principles of motion economy into four groups:

1. Principles related to the use of human body,
2. Principles related to the arrangement of the workplace,
3. Principles related to the design of tools and equipment,
4. Principles related to time conservation.

Work Organization and Methods Engineering for Productivity.
DOI: https://doi.org/10.1016/B978-0-12-819956-5.00013-3

13.3.1 Principles related to the use of human body

1. The two hands should begin motions at the same time.
2. The two hands should not be idle at the same time except during rest periods.
3. Motions of the arms should be made in opposite and symmetrical directions and should be made simultaneously.
4. Hand motions should be confined to the lowest classification with which it is possible to perform the work satisfactorily.

   These hand motions can be further categorised as
5. Momentum should be employed to assist the worker whenever possible, and it should be reduced to a minimum if it must be overcome by muscular effort.
6. Smooth continuous motions of the hands are preferable to zigzag motions or straight-line motions involving sudden and sharp changes in direction.
7. Ballistic (i.e. free swinging) movements are faster, easier, and more accurate than restricted or controlled movements.

13.3.2 Principles related to the arrangement of the workplace

1. There should be a definite and fixed place for all tools and materials.
2. Tools, materials, and controls should be located close in and directly in front of the operator.
3. Drop deliveries should be used whenever possible.
4. Materials and tools should be located to permit the best sequence of motions.
5. Arrange the height of the workplace and chair for alternate sitting and standing, when possible.
6. Provide a chair of the type and height to permit good posture with adequate lighting.
7. The color of the work place should contrast with that of the work and thus reduce eye fatigue.

Fig. 13.1 gives two illustrations of the minimum and maximum distance to which the operative can stretch the hands comfortably with least strain. While diagram alongside indicates the perspective view of the hand movement limitations, diagram below indicates the plan of the same. The work area should be so designed to locate all the tools and parts between the minimum and maximum distances to facilitate the easy picking up by the operator (Tables 13.1 and 13.2).

Fig. 13.2 illustrates the arrangement of the worktable for assembly of small components like screws, nuts and washers to be located in sliding chutes. While Fig. 13.2A shows, an optimal gravity based small component holder and feeder, as explained in point. No 6 of the principles related to the design of tools and equipment, Fig. 13.2C shows how a retractable screw driver can be located in the work area, within the reach of the operative.

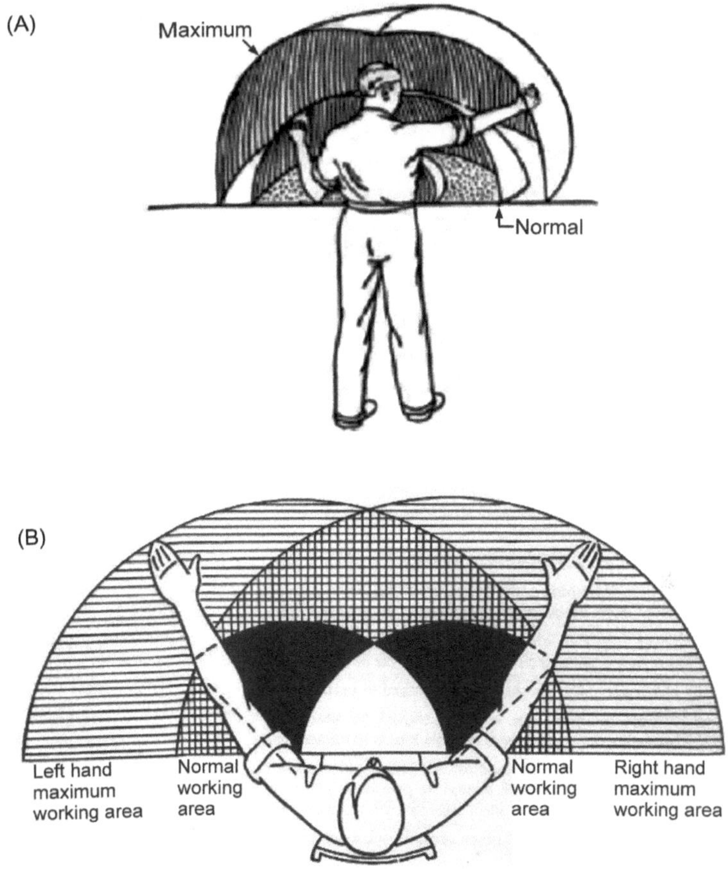

FIGURE 13.1 Maximum and minimum reach area at a worktable. *From Principles of work study - courtesy ILO.*

TABLE 13.1 Categories of hand motions.

Class	Pivot	Body members moved
1	Knuckle	Finger
2	Wrist	Hand & fingers
3	Elbow	Forearm, hand & fingers
4	Shoulder	Upper arm, forearm, hand & fingers
5	Trunk	Torso, upper arm, forearm, hand & fingers resulting in disturbance of the posture

TABLE 13.2 Effective and ineffective basic motions.

Effective	Ineffective
Reach	Hold
Move	Rest
Grasp	Position
Release load	Search
Use	Select
Assemble	Plan
Disassemble	Unavoidable delay
Pre-position	Avoidable delay
	Inspect

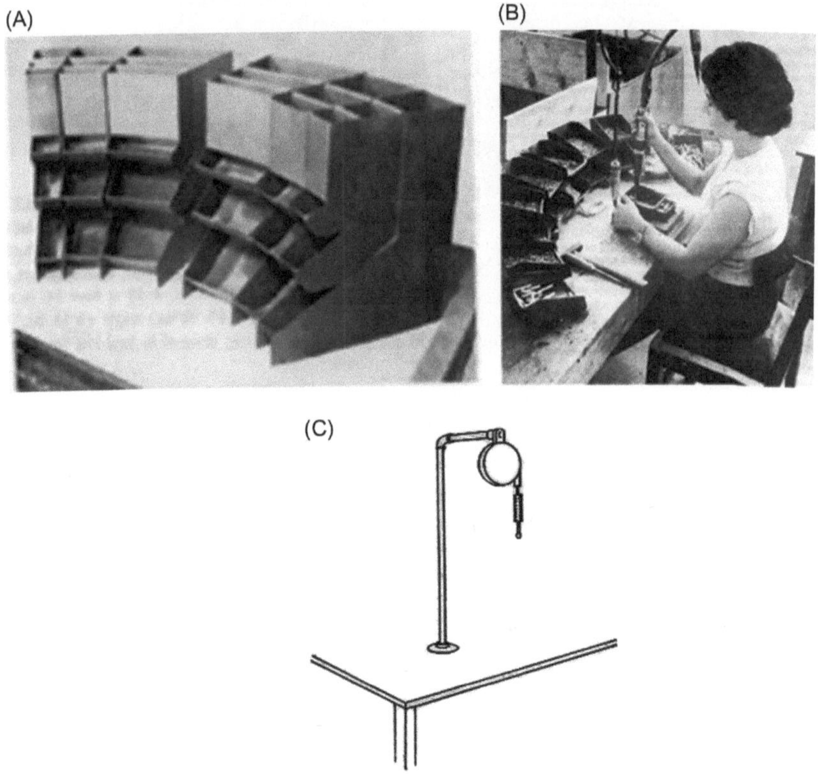

FIGURE 13.2 Optimal work area for an assembly operation. *From Principles of work study -courtesy ILO.*

The central diagram Fig. 13.2B illustrates the optimally designed work place facilitates effective assembly operation

13.3.3 Principles related to the design of tools and equipment

1. Combine tools whenever possible.
2. Preposition tools and materials.
3. Where each finger performs some specific movement like while typing, the load should be distributed in accordance with the inherent capacities of the fingers.
4. For light assembly, a screwdriver handle should be small and light.
5. Momentum should be used to help the worker in doing their task not to increase their task, like using drop deliveries or ejectors where possible.
6. Levers, switches, and other machine controls should be so placed to use them with the least change in the posture and position

The following are some of the motion economy devices which should be optimally located.

(i) Stops, which enable simultaneous working of locating and operating
(ii) quick-acting Clamps
(iii) Ball joint or Swivel fixtures
(iv) Hinged fixtures
(v) Air cylinder devices
(vi) Rotary assembly fixtures
(vii) Hoppers and motion economy bins
(viii) Removable table tops
(ix) Chutes and other drop delivery systems
(x) Foot pedals

13.3.4 Principles related to time conservation

- Even a temporary delay of work by a man or machine should not be encouraged.
- Machine should not run idle, it is not desirable that a lathe machine is running and its job is rotating but no cut is being taken.
- Two or more jobs should be worked upon at the same time or two or more operations should be carried out on a job simultaneously if possible.
- Number of motion involved in completing a job should be minimized.

13.4 Limitations of motion economy principles of Gilbreth

Motion economy principles have the following limitations.

- They do not account for physical limitations or differences in operators.

- A movement that appears ineffective from a motion economy perspective may prevent fatigue and possible injury from static posture loading, coming in way of a rationalized design procedure to ensure a productive, safe, and optimum workstation.

13.5 Therbligs

As an extension of the development of the above motion economy principles, Gilbreth identified 18 finer movements of the body. He classified and named each of them, as illustrated in Fig. 13.3. He called them a therbligs spelled in the reverse of his own name. Most of the definitions on therbligs cite Gilbreth's name as the developer of this term. Basically, Therbligs are the basic building blocks of virtually all manual work performed at a single workplace and consisting primarily of hand motions.

Originally in an article published in 1915, Frank Gilbreth wrote of 16 elements which are as illustrated in Fig. 13.3. He later added two more,

- Find (sl. No. 2) and
- Plan (sl. No 17).

Each therblig represents time and energy spent by a worker to perform a task. If the task is repetitive, of relatively short duration, and will be performed many times, it may be appropriate to analyze the therbligs that make up the work cycle as part of the work design process.

With some modification, these basic motion elements are used today in a number of work measurement systems, such as Methods Time Measurement (MTM) and the Maynard Operation Sequence Technique (MOST), which are discussed in detail in the next chapter.

13.6 Effective and ineffective classification of basic motion elements

These basic motions can be classified into 2 groups,

- The effective elements include the motions that do the primary work
- The ineffective elements include the motions that do not perform the basic work, but provide the secondary motion with a supportive role.

13.7 Objectives of therbligs

1. Eliminate ineffective therbligs if possible
2. Avoid holding objects with hand, by using work holders where possible.
3. Combine therbligs by performing right-hand and left-hand motions simultaneously
4. Simplify overall method
5. Reduce time for a motion, e.g., shorten distance

S.No	Therblig	Symbol	Abbr	Explanation	Colour
1	Search		Sh	Starts when eyes or hands or both begin to search and ends on locating.	Black
2	Find		F	Mental reaction after search	Gray
3	Select		Sl	Choosing one item among the lot	Light gray
4	Grasp		G	Taking hold of the item	Red
5	Hold		H	Prolonged grasp of the item	Gold Ochre
6	Transport unloaded		TE	Moving an empty hand for reaching an item	Olive green
7	Transport loaded		TL	Moving an item from one place to another	Green
8	Release load		RL	Releasing the item at the desired place	Carmine red
9	Assemble		A	Putting parts together into an integral part	Violet
10	Disassemble		DA	Separating one item from an assembly	Light violet
11	Use		U	Causing the device to perform its function	Purple
12	Inspect		I	Examine or test the size etc.	Burnt ochre
13	Position		P	Placing an item at a definite location	Blue
14	Pre-position		PP	Placing an item ready for use. The symbol signifies positioning in a fixture without holding	Pale blue
15	Un avoidable delay		UD	Delay beyond the control of the operator	Yellow
16	Avoidable delay		AD	Delay within the control of the operator	Lemon yellow
17	Plan		Pn	Mental action to plan how to proceed further	Brown
18	Rest		R	Take rest or pause to overcome fatigue	Orange

FIGURE 13.3 Therblig.

13.8 Some definitions of therbligs

Most of the definitions below refer to the therbligs being founded by Gilbreth, as explained in para 13.5.

> *Therbligs are some of the manual, visual, or mental elements into which an industrial manual operation may be analyzed in time and motion study.*
>
> Merriam Webster Dictionary.

> *Therbligs are 18 Kinds of Elemental Motions used in the study of Motion economy in the workplace. A workplace task is analyzed by recording each of the therblig units for a process, with the results used for optimization of manual labor by eliminating unneeded movements.*
>
> https://www.scribd.com.

> *The term therblig can be used to classify the motions of fingers, arms, hands, or the activity of the body as a whole.*
>
> http://thediagram.com.

> *Therbligs are the smallest units of work, developed by Frank Gilbreth, to develop ways to decrease unnecessary motion in several fields,*
>
> https://hubpages.com/.

> *Therbligs are a set of fundamental motions required for a worker to perform a manual operation.*
>
> Wiktionary.

> *Therblig is the smallest unit of productive motion, as analyzed by early 20th century motion engineer, Frank Gilbreth.*
>
> http://polymathsolution.com.

> *Therbligs are the unavoidable and avoidable elements of movement found by Gilbreth, in the inversion of his name Preventable delay, planning, resting and holding. The Therbligs allow the analyst to synthesize the presenting times of manual activities synthetically as a building block system.*
>
> Springer Gabler.

> *A unit of motion, at times, utilized to depict and document industrial functions for the purpose of time and movement analyzes.*
>
> Psychology Dictionary.

13.9 Conclusion

As discussed in this chapter, if the body movements are studied, analyzed and improved by the application of motion economy principles, fatigue and unnecessary movements by the worker would be reduced leading to the reduction in the work-related trauma. This would automatically improve the operative morale and increase the output in manufacturing. We have also seen how the concept of representing each motion by symbols aids the analysis and improvement.

Criteria questions

1. What do you understand by motion economy principles? (13.2)
2. Explain the broad categories of principles of motion economy. (13.3)
3. What parts of the body can be moved with trunk as the pivot? (13.3)
4. Explain what you understand by Maximum and minimum reach area at a work table. (13.3)
5. Sketch a rough layout of the table for the assembly of a screw with 2 washers and a nut, based on principles of motion economy. (13.3)
6. What are therbligs? How did they get the name? (13.5)
7. What are the objectives of therbligs? (13.5)
8. What is the difference between...... (13.5)
   a. Grasp and hold
   b. Avoidable delay and Unavoidable delay
   c. Position and Pre-position
9. Assemble and Dissemble
10. What are effective and ineffective body motions? (13.6)

Further reading

1. Gilbreth Jr. FB, Carey EG. *Cheaper by the dozen*. New York: HarperCollins; 1948.
2. ILO. *Introduction to work study*. 3rd ed. ILO; 1979.
3. Barnes RM. *Motion and time study*. 7th ed. New York: John Wiley & Sons; 1980.
4. Mundel M. *Motion and time study*. 6th ed. Prentice Hall; 1988.
5. Kiran DR. *Notes for NCES*. NPC; 1985.
6. Meyers FE, Stewart JR. *Motion and time study: for lean manufacturing*. Prentice Hall; 2002.
7. Salvendy G. *Classification of human motions*. 2004.
8. http://www.academia.edu/4932719/applicati_ons_of_principles_of_motion_economy.
9. https://en.wikipedia.org/wiki/principles_of_motion_economy.
10. www.strategosinc.com/motion_economy.htm.

Chapter 14

Work measurement

14.1 Introduction

Among the several tools of work study, work measurement, is a procedure for measuring the time taken in the performance of an operation or series of operations in such a way that ineffective time is shown up and can be separated from effective time. As the name indicates, it provides the essential data on the operational timings which is vital to the production scheduling activity for determining the machine wise and component wise operational times. It establishes the work content of the job by stopwatch time study or synthetic time standards like PMTS. In effect, it is the application of the techniques designed to establish the time for a qualifier worker to carry out a specified job at a desired level of performance. As illustrated in the Chapter 6 on Method Study, work measurement assists in identifying and quantifying any ineffective time in any operation.

The structured and systematic work measurement gives a common platform for the evaluation and comparison of all types of work. As indicated in the previous paragraph, the results obtained from work measurement are commonly used as the basis of the planning and scheduling of work, manpower planning, work balancing in team working, costing, labor performance measurement, and financial incentives. They are also used, though less commonly used as the basis of product design, methods comparison, work sequencing and workplace design.

The stopwatch time study approach for work measurement was popular and widespread during the 1970s and is ideal and accurate for jobs that are highly repetitive. In fact, in the automotive, textile, and other manufacturing industries, the productivity incentives for virtually every production job was based on standard times developed from time studies. Paragraph 14.11 explains how to fill in the time study data sheet, and Fig. 14.5 illustrates the sheet.

While Method study is concerned with reduction of the work involved, primarily by eliminating unnecessary movement on the part of material or operatives and by substituting good methods for poor ones, work measurement is concerned with investigating, reducing and subsequently eliminating ineffective time, that is time during which no effective work is being performed, whatever may be the cause. Work measurement supplements method study by uncovering non-standardized elements and non-value adding activities that exist in the workplace.

Work Organization and Methods Engineering for Productivity.
DOI: https://doi.org/10.1016/B978-0-12-819956-5.00014-5

14.2 Definitions on work measurement

Work Measurement is the application of techniques designed to establish the time for an average worker to carry out a specified manufacturing task at a defined level of performance.

Wikipedia.

Work Measurement is the application of time and motion study and activity sampling techniques to determine the time for a qualified worker to complete a specific job at a defined level of performance.

Business Dictionary.

Work Measurement is a term which covers several different ways of finding out how long a job or part of a job should take to complete. It can be defined as the systematic determination, through the use of various techniques, of the amount of effective physical and mental work in terms of work units in a specified task. The work units usually are given in standard minutes or standard hours.

http://www.managers-net.com/workmeasurement.html.

Work measurement is the process of establishing the time that a given task would take when performed by a qualified worker working at a defined level of performance.

http://www.ims-productivity.com/page.cfm/content/Work-Measurement.

As we had seen, work measurement is the determination of the length of time it should take to complete a job. As a by-product work measurement also helps to uncover non-standardization that exist in the workplace and non-value adding activities and waste. That is the reason why method study and work measurement are considered as the perfect wedded couple, though they have different procedures.

14.3 Objectives of work measurement

We can summarize the objectives of work measurement as below.

14.3.1 Comparison purpose

1. To compare the efficiency of alternative methods.
2. To facilitate the selection of ideal method/process from alternatives.
3. To check actual times against estimated times.
4. To check actual operator or machine performance against the expected values.
5. To establish the benchmark jobs for comparative purposes.

14.3.2 Capacity assessment

1. To provide information on which the planning and scheduling of production can be based, including computation of the plant and labor requirements.
2. To convert the work content of the orders to the available plant capacity.
3. To enable realistic plans to be drawn up.
4. To establish operator capacity
5. To help balancing of the workloads between operator groups, departments or plant sections in association with multiple activity charts, etc.
6. To provide perfect line balancing especially of assembly lines.
7. To determine, in association with man and machine multiple activity charts, the number of machines an operative can run.

14.3.3 Estimating purpose

1. To discover and eliminate lost or ineffective time.
2. To set operating goals and objectives.
3. To provide information on which estimates for tenders, selling prices and delivery promises can be based
4. To provide information for labor-cost control and to enable standard costs to be fixed and maintained.
5. To help in projecting the company's future plans.

14.3.4 Wage payment process

1. To establish standard times for performance measurement. However, this principle of productivity incentives has been disproved as a factor for increasing operative resistance and inhibiting quality improvement. Hence the standard time establishment should not be emphasized for operator performance but for incentive payment.
2. To provide fair returns on possible incentive bonus payment schemes.
3. To measure performance against realistic expectations.

14.4 Principal techniques for work measurement

The following are the principal techniques by which work measurement is carried out:

1. Stopwatch time study
2. Activity sampling
3. Predetermined motion time systems (PMTS)
4. Synthesis from standard data
5. Estimating
6. Analytical estimating
7. Comparative estimating

14.5 Stopwatch time study

Among those indicated in paragraph 14.4, time study forms the basic work measurement mode and discussed in detail in this chapter. It systematically records times by use of stopwatches, either traditional mechanical type or the modern electronic type. It also records the rates of performing a job or elements of a specified job to be carried out under specified conditions and at defined levels of performance. Since a stop watch has traditionally been used to measure these times, this process is called stop watch time study. Nevertheless, several improvements and modernizations have taken pace in the design and use of these timing gadgets, which are explained in paragraph 14.6., and also in later chapters.

14.6 Equipment required for stopwatch time study

1. Stopwatch
2. Observation sheet
3. Observation board
4. Stationary - Pen, Pencil, Eraser, Calculator
5. Electronic data collector and computer

14.6.1 Stopwatch

The stopwatch, which can either be

- Plain Decimal-minute mechanical stopwatch for normal industrial operations, which are generally of short duration (Fig. 14.1A).
- Decimal Hour mechanical stopwatch for long operations like ore/cargo handling in sea or airports (Fig. 14.1B).
- Split hand stop watches, where two hands move together. When the knob is pressed the main hand continues while the other hand stops to enable

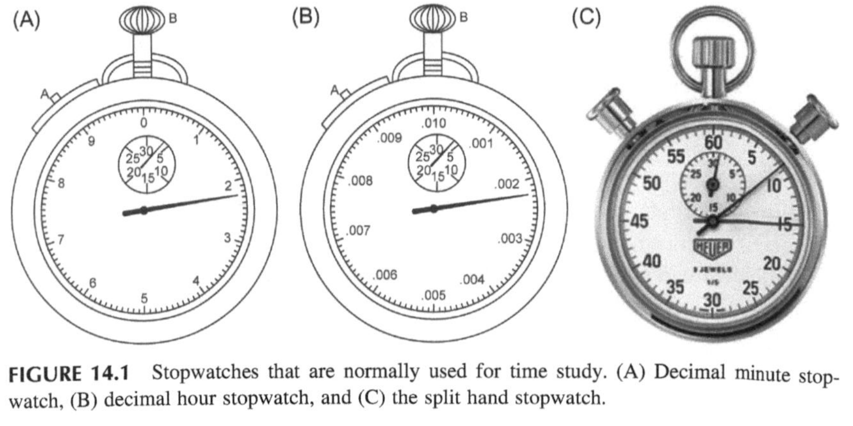

FIGURE 14.1 Stopwatches that are normally used for time study. (A) Decimal minute stopwatch, (B) decimal hour stopwatch, and (C) the split hand stopwatch.

FIGURE 14.2 Digital stopwatch.

FIGURE 14.3 How to hold a stopwatch.

the time elapsed to be recorded. Another press to the key makes the split hand join the main hand (Fig. 14.1C). A press on the side knob causes both hands to return to zero.

- Digital or electronics stopwatches, similar to the digital wrist watches (Fig. 14.2).

Fig. 14.3 shows, how to hold a stopwatch while timing, unless it is fixed in to the observation board. For example, while timing in a running race.

14.6.2 Observation sheet

This is also called the recording sheet, as illustrated in Fig. 14.5 and explained further in paragraph 14.12.

14.6.3 Observation board

This is nothing but a small plywood board with a spring clip to hold the recording sheet in place while recording. The stopwatch is fixed on to the observation board (Fig. 14.4A) to enable simultaneous holding, pressing and recording. The board is shaped suitable to fit the waist while recording.

Sometimes instead of a single split hand watch, three normal stopwatches are fixed on to the board, one in the zero position, the second in the running position, and the third in the stop position and all the three can be controlled by single lever as in Fig. 14.4B.

14.6.4 Other equipment used

Cameras

In the recent times cameras and computers are increasingly being used for timing the operations. Paragraphs 7.16 to 7.19 of the Chapter 7 on Record, illustrates such uses of cameras. They illustrate the motion picture camera running at a constant speed, the film being projected at an equal constant speed. The most recent to be added is the time lapse camera video, useful for analyzing large open fields operations, as illustrated in paragraph 7.20.

Time study machine

A time study machine is developed in which lines are made on a paper tape running at constant speed by pressure of the fingers on two keys. Its only advantage over the stopwatch is that it leaves the work study man free to observe the operation continuously instead of having to look at and read

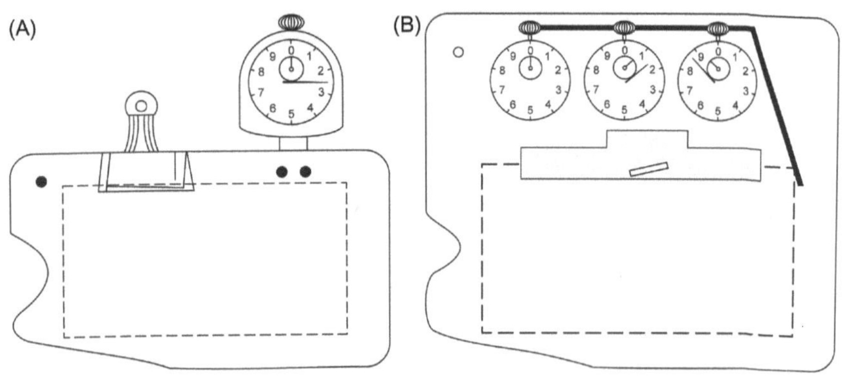

FIGURE 14.4 Observation board. (A) With a single stopwatch and (B) with three stopwatches.

the watch. It also enables very short elements to be timed. The tape has to be measured on completing the study.

Use of computers

Similarly, use of computers and computer software are now increasingly being used for time study and standard time computation. These are explained further in paragraph 14.16.

14.7 Methods of stopwatch timing

There are basically two methods of timing using a stopwatch. They are

- Fly back or Snap back method and
- Continuous or Cumulative method.

1. **Fly back method:** When two elements are to be timed and the second element is immediately after the first element, fly back stop watch is used. Here the stopwatch is started at the beginning of the first element. At the end of the element the reading is noted in the study sheet. At the same time, the side knob of the stopwatch is pressed when hand is snapped back to zero and then continues to move from zero for timing the next element. Thus each element is timed separately.
2. **Continuous reading method:** Here the stopwatch when started at the beginning of the first element and runs continuously throughout the study. At the end of each element the watch readings are recorded on the study sheet. The time for each element is calculated by successive subtraction. The final reading of the stopwatch gives the total time known as observed time. At the end of the cycle, the knob is pressed to stop the long hand wherever it is. A third press will bring the hand to zero position, when a fresh cycle can be timed.
3. **Split hand stopwatch:** While each of the above two have their pros and cons, a split hand stopwatch can be used for a combined advantage. This has 3 knobs and 2 hands. On the first press, both hands start moving together. When the first element is ended, a second press would stop one hand while the other hand continues moving. This will give us enough time to record the elapsed time. On a third press, the split hand moves forward to join the other hand and the second element continues to get recorded. At the end of the cycle the central knob is pressed to bring both hands to zero.

14.7.1 Pros and cons of the 2 methods

Pros of fly back method

1. Watch readings at the end of each element are the actual elemental times
2. Less calculation as no subtraction is required in this method.

Cons of fly back method

1. There may be a time lapse between the end of an element and flying back, resulting in wrong timing.
2. If an element is missed, there is no means by which it can be taken into account afterwards.
3. This method requires a lot of skill as two things are to be done by analyst at the same time i.e., recording the time and pressing the stem knob.

Pros of continuous method

1. There is no time lost by delayed reading of the end of one element. At the most the reading may be few units more or less, which could be corrected by averaging 10 cycles.
2. Even Missing of any element or occasional element will not affect the overall time.
3. less training required for the observer.
4. As there is no possibility for altering times in the favor of management or omitting any activity or element, it is more preferred by Trade Unions.
5. This timing method is most accurate of all the timing methods as there is no scope of time lag or time lead.

Cons of continuous method

1. Additional computations needed to achieve the individual element times, resulting in increased time consumption.

14.7.2 The requirements for effective time study are

1. Defined job
2. Defined method
3. Elemental breakdown and defined break points
4. Method of timing
5. Correct equipment and recording media
6. Quality standard and checks
7. Method of assessing relative performance
8. Experienced, qualified and motivated worker
9. Co-operation and goodwill from supervisors and workforce.

Of the above, the third point viz Elemental breakdown or the division of an operation into basic elements, is the most critical. Elements should be easily identifiable, with definite beginnings and endings so that, once established, they can be repeatedly recognized. They should be as short as can be conveniently timed by the observer. These points are known as the break points and should be clearly described on the study sheet.

14.8 Elemental breakdown

In Chapter 8, paragraph 8.5.1, we have seen how splitting each and every problem into components and dealing with each sub-problem individually, helps method study analysis. Similarly dividing each operation into basic elements helps us to assess the rate of working more effectively than when the assessment is done over the entire operation.

- A work cycle is defined as the sequence of elements which are required to perform a job or yield a unit of production. The sequence may sometimes include occasional elements.
- An element is defined as the distinct part of a specified job selected for convenience of observation, measurement and analysis.

14.8.1 Objectives for the elemental identification

I.L.O.'s Introduction to Work study indicates the following objectives as well as guidelines for the identification of the elements

1. To separate productive part of task that is the effective time, from the unproductive one, causing ineffective time.
2. To ensure that rate of working is assessed more meticulously so as to improve accuracy in rating. An average worker tends to perform some elements faster and some slower. Breaking of task into short elements allows us to rate of each element separately which is more realistic than just rating once for the complete cycle.
3. To identify and isolate elements causing high fatigue. Breaking of task into short elements permits giving appropriate rest allowances to different elements.
4. To ensure that the omissions and additions of unspecified elements are detected quickly.
5. To have detailed job specifications. This helps in detection of any variation in the method that may occur after the time standard is established.
6. To set time values of very frequently occurring elements like operating machine controls, or loading and unloading work pieces so as to develop standard data for such elements.

14.8.2 Guidelines for breaking an operation into elements

1. Elements should be easily identified and should have distinct beginning and end.
2. The break points should be clearly described by breaking the job into elements and recoding them on time study charts.

3. Elements should not be too small in time variations (under 3 s) nor too large (20 s). If they are too small, stopwatch reading and recoding would be difficult. If they are too large then the performing rate may vary and making it difficult to rate it properly.
4. The above point brings out the fact that the basic therbligs like Grasp, hold, position propounded by Gilbreth, may be too small to be timed as elements for normal time study.
5. Each element must be measured in terms of time units. Decimal minute units are preferred than minutes and seconds.

14.8.3 Types of elements

The British Standard Glossary of Terms in Work Study, as cited by ILO in the book Introduction to Work study (2nd ed.) lists 8 types of elements as explained below, together with examples.

1. Repetitive element
2. Occasional element
3. Constant element
4. Variable element
5. Manual element
6. Machine element
7. Governing element
8. Foreign element

The following examples will help us in identifying the type of each element we come across during time study.

1. **Repetitive element**

   A repetitive element is an element which occurs in every work cycle of the job.

   For example:
   - Change the turret head direction and advance the tool;
   - Picking up a part prior to an assembly operation;
   - Locating a workpiece in a holding device;
   - Putting aside a finished component or assembly.
2. **Occasional element**

   An occasional element is an element which does not occur in every work cycle of the job, but which may occur at regular or irregular intervals. Even though it occurs very occasionally and not in every cycle, it is a necessary and useful part of the job. Hence it has to be incorporated in the final standard time for the job.

   For example:
   - Clear swarf;
   - Adjust belt tension;
   - Tighten tool post bolts;

- Machine setting;
- Receive instructions from the foreman.

3. **Constant element**

   A constant element is an element for which the basic time remains constant whenever it is performed.

   For example:
   - Switch on machine;
   - Gauge diameter;
   - Screw on and tighten nut;
   - Insert a particular cutting tool into machine.

4. **Variable element**

   A variable element is an element for which the basic time varies in relation to some characteristics of the product, equipment or process, dimensions, weight, quality, etc.

   For example:
   - Fettle (grinding the rough projections in castings)
   - Saw logs with handsaw (time varies with hardness and diameter);
   - Sweep floor (varies with area);
   - Push trolley of parts to next shop (varies with distance).

5. **Manual element**

   A manual element is an element performed by a worker.

6. **Machine element**

   A machine element is an element automatically performed by a power-driven machine (or process).

   For example:
   - Drilling or turning operations;
   - Anneal tubes;
   - Fire tiles;
   - Form glass bottles;
   - Press car body shell to shape;

7. **Governing element**

   A governing element is an element occupying a longer time than that of any other element which is being performed concurrently.

   For example:
   - Turn diameter on a lathe, while gauging from time to time;
   - Boil kettle of water, while setting out teapot and cups;
   - Develop photographic negative, while agitating the solution occasionally.

8. **Foreign element**

   A foreign element is an element observed during a study which, after analysis, is not found to be a necessary part of the job.

   For example:
   - Sand the edge of a board before planing has been completed in furniture manufacture;
   - Degrease a part that has still to be machined further.

14.9 Number of cycles to be timed

The following general principles govern the number of cycles to get the representative average cycle time.

- Greater the accuracy desired in the results, larger should be the number of cycles observed.
- The study should be continued through sufficient number of cycles so that occasional elements such as setting-up machine, cleaning of machine or sharpening of tool are observed for a good number of times.
- Where more than one operator is doing the same job, short study (say 10–15 cycles) should be conducted on each of the several operators than one long study on a single operator.

14.10 Performance rating

As can be expected, when a certain operation is being timed, the operator sometimes performs in a deliberately slow speed to inflate the standard operational time so as to get a higher incentive later. Hence the observer assesses the speed and effectiveness of the operation in relation to his concept of Standard Rating. This is called Performance Rating as defined by BS 3138 (1979):

> *The Performance Rating corresponding to the average rate at which qualified workers will naturally work, provided that they adhere to the specified method and that they are motivated to apply themselves to their work. If the standard rating is consistently maintained and the appropriate relaxation is taken, a qualified worker will achieve standard performance over the working day or shift.*
>
> BS 3138 (1979).

An illustration of the performance rating in a standard scale:

Ask someone to walk across the room you both are in. Observe and rate his speed of walking. This, in simulation to the movement of the operator, is similar to the rating of an operator at his work, as illustrated in Table 14.1.

TABLE 14.1 Rating of walking pace.

	Activity	Average walking speed	Rating
1	No activity	0 kmph	*0*
2	Very slow stroll	2 kmph	*50*
3	Strolling	3 kmph	*75*
4	Normal walk (standard rating)	5 kmph	*100*
5	Fast	6 kmph	*125*
6	Very fast	7 kmph	*150*

14.11 Time study data sheet

Having had the basic understanding on the stopwatch reading, elemental identification and performance rating, let us now understand how to fill in the time study data sheet shown in Fig. 14.5 (See after page 191). The elements are listed vertically and the cycles horizontally. Each element for each cycle is divided into 4 boxes. While timing by continuous reading the time at the end of the element is recorded in the top left box while the performance rating is filled in the top right box. An explanatory block at the bottom center of the sheet tells what to fill where. After timing all the cycles, the observer returns to the office and fills in the other two boxes, viz the elapsed as the difference between the successive readings, and the normal time as a product of elapsed time and the rating.

14.12 Operational standard times

The operational standard time of a job or operation is defined as the total time in which a job should be completed at standard performance. It is given by the sum of the basic time + relaxation allowance + any allowance for additional work go be performed during the course of the work.

As discussed in the previous paragraphs, the basic steps involved in determining the operational standard time by a time study can be summarized as,

1. Select the operation that has to be timed.
2. First establish the method to be adapted, by observation and preliminary method study, to reduce the number of subsequent revisions and restudies.

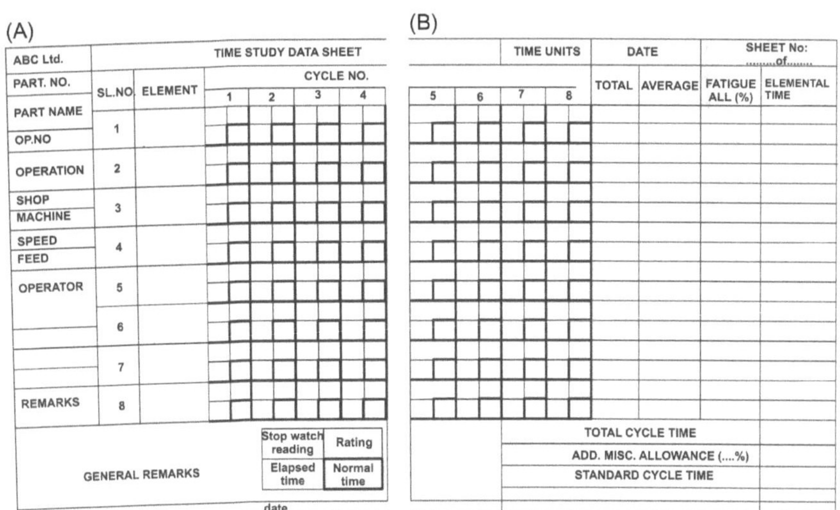

FIGURE 14.5 Time study data sheet.

3. Break down the job into short, elemental tasks with distinct "break points" between them. A break point is that point of the operation which can be distinctly identified by the observer to ensure correct and unambiguous timing.
4. Time the element. Take a position near the worker and record the stopwatch time reading noted at each break point time, on the time study observation chart illustrated in Fig. 14.5. This timing shall be done for several cycles of the operation to get an average observed element time.
5. Simultaneously rate the worker's performance. The rating factors usually range between 80% and 120%. A performance rating factor of 100% reflects normal work performance, below 100% represents a below-average performance, and above 100% indicates performance better than normal. The objective of the rating factor is to determine a normal or average elemental time for the job irrespective of the performance speed of the operator. Since the rating factor is mostly a subjective judgment, work measurement trainees would be trained to judge the normal speed and the rating factor during their on-the-job training.
6. Calculate the observed elemental time. Once a sufficient number of job cycles have been observed, an average time for each work element is calculated after coming to the work table. While such time studies were conducted during the seventies using a stopwatch attached to a clipboard today the hand-held electronic time-study machines store elemental times in a memory that can be transferred to a computer for processing. Even video cameras can be used to videotape jobs, with the time study conducted outside the workplace at a later time.
7. Compute the normal time by multiplying the elemental average time by the performance rating factor. The normal cycle time (NT) can now computed by summing the elemental normal times.
8. Compute the operational standard time. The standard time is computed by adjusting the normal cycle time by an allowance factor for unavoidable work delays (such as a machine breakdown), personal delays (such as using the restroom), and normal mental or physical fatigue as detailed in 14.12.9 and listed in Table 14.2. The allowance factor is a percentage increase in the normal cycle time. The standard time is calculated by the formula given in paragraph 14.12.1.
9. Allowances: It cannot be expected that a worker will work all day without any interruptions. So over and above the normal time, some extra time should be allowed both for the personal needs of the operator or as required by the operation or work place. These allowances, which vary for men operatives and women operatives, could be classified into
   - Personal allowance: Time for attending personal needs must be allowed which is normally 5–7% of the normal time.
   - Fatigue allowance: Sometimes the job may involve heavy physical exertions and are done under adverse conditions like heat and

TABLE 14.2 ILO recommended allowances.

	Allowance as percentage	Men	Women
1	Personal allowance	5	7
2	Basic fatigue allowance	4	4
3	Additional for standing posture	2	4
4a	Abnormal posture – slightly awkward	0	1
4b	Abnormal posture –awkward	2	3
4c	Abnormal posture – very awkward	7	7
5a	Weight lifting – 2.5 kg	0	1
5b	Weight lifting – 5 kg	1	2
5c	Weight lifting – 10 kg	3	4
5d	Weight lifting – 20 kg	10	15
5e	Weight lifting – 30 kg	19	Not allowed
5f	Weight lifting – 50 kg, (In rare cases, when no other help is available.)	58	Not allowed
6a	Well-lit work place	0	0
6b	Slightly dark	2	2
6c	Poorly lit	5	5
7a	Well ventilated. With fresh air	0	0
7b	Badly ventilated, but not hot or no fumes	5	5
7c	Close to furnaces	5	15
8a	Fairly fine work, with less eye strain	0	0
8b	Intricate work, with less eye strain	2	2
8c	Very intricate work, with high eye strain	5	5
9a	Continuous moderate noise	0	0
9b	Intermittent, loud noise	2	2
9c	High pitched high decibel noise	5	5
10a	Fairly complex process	1	1
10b	Wide span of attention	4	4
10c	Highly complex controls	8	8
11a	Mental monotony - Low	0	0
11b	Mental monotony – Medium	1	1
11c	Mental monotony - High	4	4
12a	Physical monotony - Low	0	0
12b	Physical monotony – Medium	2	1
12c	Physical monotony - High	5	2

humidity. These jobs cause fatigue and a suitable fatigue allowance as a percentage over normal time is given. This is normally determined by discussing with the supervisor and the management.

- Delay Allowance: This is also called the contingency allowance and includes delays caused by minor adjustments, tool breakages, material variations, or interruptions caused by supervisors must be included in the time standard.
  - Avoidable delays like intentional or leaving the workplace over and above the time permitted are not allowed.
  - In the case of booked equipment breakdown, repairs etc., when the worker is taken off the job, these delays are not considered. In any case maintenance and repair work is booked under machine not available time (MNA) and deducted from the total working time, and hence is outside the time study purview.

10. From steps 1–8 and after adding the predetermined allowances as given as above, the standard operational time can be calculated as in paragraph 14.12.1.

ILO recommends allowances, as illustrated in Table 14.2, to be added to the normal time.

14.12.1 Terms used in determining the operational standard time

The following are some of the terms used in determining the operational Standard time

OT = Observed operational time (observed elemental times)
BT = Basic elemental time
ST = Standard operation time
RF = Rating factor
FA = Fatigue allowance, which normally is 15% and
MA = Miscellaneous allowance which normally is 5%
And the standard time ST given by
$ST = (OT \times RF)(1 + FA + MA)$

14.13 Operational budgeted time

While standard times are used for machine scheduling, additional %age allowance is given over and above the standard times to compute the budget times, which information is used by the costing department to compute product cost and budget allocation. Fig. 14.6 illustrates how this information is converted for machine scheduling and other production planning activities.

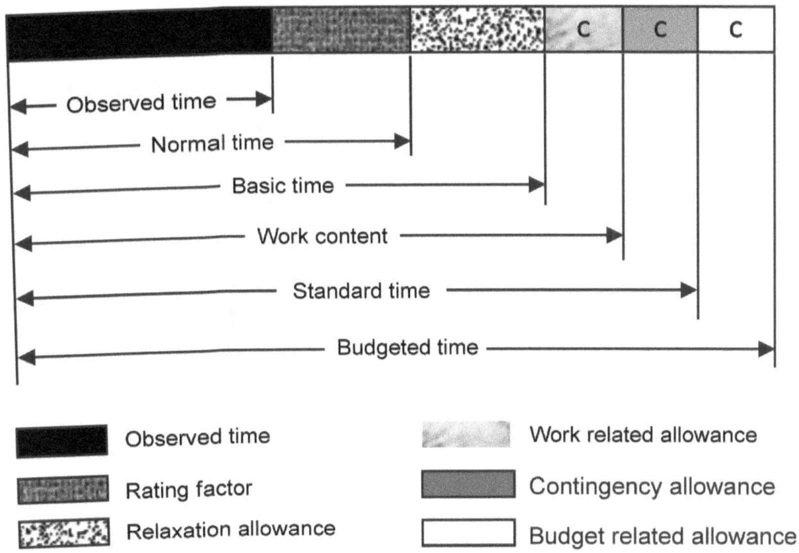

FIGURE 14.6 How operational time and budget time are computed.

14.14 Standard time declaration form

The standard time for each operation of each component is the basic data used for computing the total time each machine would be working on a particular operation of a batch of a component and this is the primary information needed for scheduling each machine. This is called Standard Time Declaration Form (Fig. 14.7). In large organizations with separate industrial engineering department this would be set by the industrial engineer and the information forwarded to the planning department, which converts this into machine wise and component wise operational times to be used for machine scheduling. The industrial engineers must understand how their work measurement tool provides the basic input information for scheduling. Readers are advised to refer to the book on Production Planning and Control, a Comprehensive Approach by this author, to understand how this standard declaration form helps in production scheduling.

14.15 Method improvement is a continuous process

As per the very principle of Kaizen there is always a scope for improving the existing method or even the previously improved method. Hence periodic method improvement studies and other cost reductions studies should be made on a periodic basis. Such reports may be submitted every month but the standard times can be revised, may be on an annual basis.

STANDARD DECLARATION FORM

NEW / REVISION / TENTATIVE

STANDARD NO.__________

DATE .______________

DEPARTMENT____________________ PRODUCT ____________________

COMPONENT____________________ COMPONENT NO.____________________

OPERATION____________________ OPERATION NO.____________________

SPEED ________________ FEED ______________________________

MACHINE____

STANDARD TOOLS____

SPECIAL TOOLS ____

GAGES____

INSPECTION BY OPERATOR __________BY INSPECTOR ____________________

STANDARD METHODS OF OPERATION

No	Operation and element	Normal time	%age allowance	Standard time

Total standard time of operation________________

Output per hour ______________per shift________________

Cc : Production / Planning / SQC / Costing / IED

Note: Any changes in the above shall immediately be brought to the notice of the Industrial Engineering Department.

FIGURE 14.7 Standard declaration form.

14.16 Computer software for work measurement

Several software for work measurement have been developed that can be loaded onto the smartphones, so that data collection on a mobile device would be more than 50% faster than using a stopwatch. This also would eliminate the step of manual data entry, and minimize the risk of potential errors.

1. Work Study + developed by Quetech Ltd is great for simple time studies, it is built to handle complex studies with 1000s of elements. It works well for slow changing processes, but is equally effective at accurately recording fast changing activities.
2. UmtPlus developed by Rapid Modeling Corporation includes 3 comprehensive software modules:
   - UmtManager: This program allows you to easily create, edit and manage any number of configurations from a PC to perform your time study or work sampling.
   - UmtPlus (for PDA): This program allows you to accurately collect data for work measurement on handheld computers.
   - StatUmt: This program allows you to instantly review reports and graphs while easily manipulating data for analysis on a PC
3. StandardsPro Work Measurement Software developed by HB Maynard Associates, is a complete engineered standards development tool designed specifically to create, maintain, retrieve, and analyze time standards for all types of work. You can create individual, measurable units of work that describe short-cycle and highly repetitive tasks as well as data for work that is long-cycled and unique. By using a hierarchical data structure, these individual units of work can then be combined into standards and eventually into plans for parts, components and products.
4. AccuStudy, as published in the June 2018 edition of the e-newsletter of Modern Machine Shop magazine, is a software program developed to run on a variety of handheld, touch screen computers. Many of these computers are ideally suited for industrial environments and use Windows-CE as the operating system.
5. MTM-1: Methods-Time Measurement (MTM-1), which is explained further in the next chapter, is a procedure which analyzes any manual operation or method into the Basic Motions required to perform it, and assigns to each motion a predetermined time standard, which is determined by the Influencing Factors under which it is made. MTM-1 is used for Process Type 1 activities, where the following application conditions exist:
   - production with a very high degree of repetition
   - short cycle work processes
   - detail-designed processes
   - work instructions with precise descriptions of methods
   - deviations in individual operator methods must be minimized
6. MOST (Maynard Operation Sequence Technique) is an activity based work measurement system that enables us to calculate the length of time required to perform a task, as further explained in the next chapter.

14.17 Conclusion

Work measurement is the basis for providing the operational time standards in a workshop. As indicated in paragraph 14.3, these operational standard times provide the basic information, especially for the Production Planning and Control engineer to perform his routing and scheduling functions effectively.

Criteria questions

1. How do you distinguish between time study and work measurement? (14.1)
2. What are the objectives of Work Measurement? (14.3)
3. Enumerate the different techniques of Work Measurement. (14.3)
4. What are the types of stopwatches used for time study? (14.4)
5. Why should be the observation board so shaped? (14.6)
6. What is the basic difference between fly back method and continuous reading method of time study? (14.7)
7. What does a split hand stopwatch? How does it work? (14.7)
8. Compare the pros and cons of fly back and continuous reading methods. (14.7)
9. What is meant by Elemental breakdown and how do you identify the elements? (14.8)
10. List and discuss the several types of elements. (14.8)
11. What is performance rating? Illustrate how you judge that of an operative. (14.10)
12. Draw and explain the details of a time study data sheet. (14.11)
13. What is operational standard time? How do you determine it? (14.12)
14. What are allowances and why they are added to the observed time? (14.13)
15. List some of the computer software available for work measurement. How does StandardsPro software work? (14.14)

Further reading

1. International Labor Organization. *Introduction to work study*. 3rd ed. 1979.
2. Kiran DR. *Production planning and control: a comprehensive approach*. BS Publications; 2018.
3. Groover MP. *Work systems: the methods, measurement & management of work*. Prentice Hall; 2007. ISBN 978-0-13–140650–6 h.
4. http://www.laubrass.com/umtplus.
5. http://www.rapidmodeling.com/work-measurement-software.
6. http://www.quetech.com/.
7. https://www.hbmaynard.com/softwareproducts/StandardsPro.asp.

Chapter 15

Micro motion study

15.1 Introduction

In the previous chapter, we have seen how time study is the basic work measurement methodology for setting up work standards. In Chapter 13 we have also seen how the study and analysis of the basic human body motions, play a significant role in eliminating unwanted voluntary movements of the operative, resulting not only in overall time reduction and cost saving, but also in reducing the effort and fatigue experienced by the operative during the course of work. Gilbreth identified these basic human body motions and called them therbligs. Though they were too small to be measured by stopwatches, substantial work has later been conducted on these motions in computing and establishing the time standards for each motion in terms of certain parameters. These are compiled from a very large number of studies of each movement, generally by a frame-by-frame analysis of films of several operatives, both men and women, performing a wide variety of tasks. Micro Motion Study involves detailed study of these movements either by direct observations or by filming and studying the movements, as in chronocyclographs, detailed in Chapter 7.

Micro motion study is defined as the technique of recording and analyzing the timing of basic elements of an operation with the objective of achieving the best method of performing the operation. It is best suited for those operations or activities which are of short duration and which are repeated hundreds of times.

15.2 Predetermined motion time standards

The standard times developed for these micro-motions are called predetermined motion time standards and the method of computing them is called Predetermined Motion Time System (PMTS). The main purpose of PMTS lies in the estimation of time for the performance of a task by means of observing and timing the fine body movements. While PMTS is the common name given for this group of methodologies, as we will see in subsequent paragraphs, some definitions offered for PMTS as a group, are given below

Work Organization and Methods Engineering for Productivity.
DOI: https://doi.org/10.1016/B978-0-12-819956-5.00015-7

Definitions on PMTS

PMTS refers to the tables of time data at defined rates of working for classified human movements and mental activities. Times for an operation or task are derived using precise conventions. Predetermined motion time data have also been developed for common combinations of basic human movements and mental activities.

BS 3138, Glossary of Terms Used in Work Study.

Predetermined Motion Time Systems (PMTS) are work measurement systems based on the analysis of work into basic human movements, classified according to the nature of each movement and the conditions under which it is made. Tables of data provide a time, at a defined rate of working, for each classification of each movement.

Institute of Management Services.

The concept of PMTS is to analyze a job into its fundamental human activities, apply basic times for these from tables and synthesize them into a basic time for the complete job.

http://www.managers-net.com/pmts.html.

PMTS can be defined as a procedure that analyzes any manual activity in terms of basic or fundamental motions required to perform it. Each of these motions is assigned a previously established standard time value and then the timings for the individual motions are synthesized to obtain the total time needed for performing the activity.

http://nptel.ac.in/courses/112107142/part1/lecture11.htm.

Predetermined Motion Time System (PMTS) consists of a set of time data for basic human motions and a systematic procedure which analyzes and subdivides any manual operation of human task into motions, body movements, or other elements of human performance, and assigns to each the appropriate time value.

Business and Management University, Graduate School of Management.

15.3 Objectives of PMTS

1. Developing time standards for fine body movements.
2. Comparing the times for alternative proposed methods and also the economics of the proposals prior to production run.

3. Estimating the requirement of manpower, equipment and space requirements prior to setting up the facilities and start of production.
4. Developing tentative work layouts for assembly line for minimizing the subsequent re-arrangement.
5. Achieving optimal line balancing especially in assembly lines.
6. Verifying the results of direct time study.

15.4 Advantages and limitations of PMT systems

15.4.1 Advantages

Compared to other work measurement techniques, all PMT systems claim the following advantages:

1. No stopwatch time study is needed.
2. From the drawings itself, the basic motions can be identified, their parameters like angle of rotation and distance moved (mostly in cm), and the time for the operations can be estimated before the job is actually done. Thus PMTS helps us in estimating the operational elements during the product design stage itself.
3. The elemental standard times established by PMTS can be usefully applied to several jobs done on various types of machines and equipment.
4. This is very useful in production planning, forecasting, equipment selection, etc.
5. The troublesome and controversial performance rating can be avoided by PMTS and many companies have preferred to use this technique.
6. The analysis done during PMTS itself would enable us to develop alternative methods resulting in further improvement in the method.
7. PMTS helps us in developing standard operation procedures (SOP) to set a detailed record of the method of operation. This helps in the operative training and method installation,
8. It also aids in the detection and verification of any future changes in the method of operation.
9. The basic times determined with the use of PMT system are relatively more consistent.
10. The development of software and extensive use of computers have made PMTS more practical.

15.4.2 Limitations

1. It can be applied to only manual elements of job and not machine elements.
2. Needs highly trained personnel.

3. Even though rating is not needed for PMTS, Some judgment is still has to be exercised at different stages.

15.5 Categories of PMTS

Normally PMT Systems can be categories into 3 generations depending upon the accuracy of results required.

The first sub-system, originally termed as Methods-Time Measurement (MTM), later called MTM1 or the first generation, to distinguish from other developments or generations. MTM1 uses the basic motions and is highly accurate.

However, MTM1 is highly detailed, requires too much analysis for producing extremely accurate results. While its benefit is the accuracy achieved, its disadvantage is that it is too much time consuming, and in practice, we may not need such highly accurate and detailed analysis but want quick results.

Second generation, termed as MTM2, is moderately accurate and also called as:

- Simplified PMTS,
- Master Standard Data,
- Primary Standard Data,
- MTM2.

Third generation called MTM3-where quick estimation of times with an accuracy of 10% is needed, and MMMM (4 M) are such extension for easier estimation, but now they are rarely adapted for work measurement and are almost obsolete.

15.6 Methods-times measurement

As discussed above, Methods-Times Measurement (MTM) is a family of PMTS systems operating at different levels called generations as listed above, and applicable to different types of work. MTM1, developed in the 1940 s by JL Schwab, GJ Stegemerten et al., during a consultancy assignment at the Westinghouse, is the first generation or the highest-level, and is the most detailed MTM obtained by analyzing large numbers of repetitive cycles of manual work on film. This system quickly became popular in USA and Canada that MTM Association for Standards and Research was formed in 1951 in USA and an International MTM Directorate (IMD) was formed on paris in 1957.

Here the basic motions of Reach and Grasp are treated as separate basic movements. A table can be made indicating the times taken to reach a certain object in proportion to the distance. Similarly, the time taken to Grasp is given in a table proportional to the size and weight of the item. Similar tables give timing for each of the other basic movements categorised and measured in a similar manner. MTM is suitable for measuring short cycle, highly-repetitive work.

The rating system considers four following factors independently:

- Skill -Proficiency in following the given method
- Effort -The will to work
- Conditions-The general work surroundings
- Consistency-of performance

15.7 MTM2

On the other hand, in the second generation of MTM, termed as MTM2, the motions of Reach and Grasp are combined into a composite motion, GET. Here the high level of discriminating individual therbligs like MTM1 is not needed in terms of accuracy requirements. It is hence quicker to apply, and is more suitable for longer-cycle work. It was developed by the International Standing Committee for Applied Research in 1964 and is applied for analysis of

- production with a high degree of repetition
- longer-cycle work processes
- detail-designed processes
- work instructions with precise descriptions of methods.

15.8 Maynard operation sequence technique (MOST)

MOST was developed in 1972 by H.B. Maynard and Company Inc, New York of H.B. Maynard, a leading management consultant, more known for his Handbook on Industrial Engineering, Like other MTM systems, MOST breaks down each task into individual therblings, and each is assigned a numerical time value in TMU. By analyzing the motion sequences as defined by MOST standard, it assigns TIMUs to each of the parameter of the method sequence for each task.

All the TMUs are then added together with the required allowances to arrive at the standard time, the difference lying only in the assessment of the allowances and the method of computation. While the normal MTM is a global standard, MOST is more popular in Asian countries.

MOST Work Measurement Systems are further subdivided into

(i) MOST Application Systems,
(ii) MOST Computer Systems,
(iii) Basic MOST - Mini MOST,
(iv) Maxi MOST,
(v) Clerical MOST, which is subsequently renamed as AdminMOST.

While the detailed description of each of the above method is beyond the scope of this book, reference may be made to the book, MOST Work Measurement Systems, authored by Zandin, K.B. But in a nutshell we can say that the MTM1, MiniMOST, etc. are more accurate and with complex computations, applicable to small cycle tasks, while MTM2, MaxiMOST etc. are less complex and applicable to larger cycle tasks.

15.9 Benefits of MTM systems

The benefits of MTM systems can be briefed as below,

1. Developing effective work methods prior to production
2. Improving existing methods to increase production and decrease labor cost per unit
3. Establishing time standards as basis for wages and incentive plans
4. Developing time formulas or standard data for future use
5. Guiding product design, developing effective tool designs and selecting effective equipment for most efficient manufacture

15.10 Time measurement unit

The base unit for time is the second, whether in FPS system or in CGS system. But for normal work measurement it is the minute, while for PMTS the time unit adapted is TMU (Time Measurement Unit), with 100,000TMUs in an hour, meaning each TMU measures 36 ms, or 0.036 s. These smaller units allow for more accurate calculations without the use of decimals. For example, the elements 'Toss' and 'simple pickup' may hardly take 0.1 or 0.06 s, which makes computing difficult with so many decimal points. However, in TMU's they become 3 and 2 TMUs respectively, and you can understand how the individual additions and computations are made easy.

15.11 Conclusion

The evolution of PMTS systems can be chronicled as below,

1. Time Study
2. Motion Study
3. Time and Motion Study
4. Predetermined Motion Time Systems (PMTS) Methods
5. Methods Time Measurement (MTM)
6. Methods-Time Measurement-2 (MTM-2)
7. Methods-Time Measurement-3 (MTM-3)
8. Maynard Operation Sequence Technique (MOST)

Criteria questions

1. Compare micro motion study with work measurement. (15.1)
2. What is PMTS and where is it used? (15.2)
3. Enumerate the benefits of PMTS. (15.4)
4. What do you understand by first generation and second generation PMTS? (15.5 to 15.8)
5. What is MOST and what are its types? (15.9)
6. Discuss what you know about TMU. (15.11)

Further reading

1. ILO. *Introduction to work study*. 3rd ed. ILO; 1979.
2. Maynard HB, Stegemerten GJ, Schwab JL. *Methods-time measurement.* 1948.
3. Zandin KB. *MOST work measurement systems*. 2nd ed. Marcel Dekker; 1980.
4. https://en.wikipedia.org/wiki/Predetermined_motion_time_system.
5. www.managers-net.com/pmts.html.
6. www.pmcorp.com/Services/ConsultingService.
7. https://museum-madness.blogspot.com.
8. www.tsijournals.com/articles/time-analysis-with-most-technique.pdf.
9. www.iise.org/uploadedFiles/Webcasts/Members.

Chapter 16

Ergonomics and work study

16.1 Introduction

In the earlier chapters, we have seen how method study involves the study of the methods of operations to reduce the operational times. Similarly, ergonomics involves the study of the operations with reference to the working conditions to reduce the strain in the body of the operator. It is a scientific study of people at work and the workplace with reference to the movement of the human body. It looks at what kind of work that is done, the tools used and the whole job environment including the posture of the body, the movements necessitating the human effort and the physical strain the operator is subjected to.

Ergonomics is concerned with many disciplines in its study of humans and their environments, including

(i) Anthropometry,
(ii) Biomechanics,
(iii) Kinesiology,
(iv) Physiology,
(v) Cognitive psychology,
(vi) Industrial and organizational psychology, and
(vii) Space psychology.

and this knowledge is very much essential for the engineers, especially, industrial engineers, Industrial designers, and information designers.

16.2 Aims of ergonomics

1. The goal of ergonomics is to make the job to suit the operator with lesser muscle fatigue, and to reduce the number and severity of work-related Musculoskeletal Disorders (MSDs), caused by stress as well as injuries and disorders associated with the overuse of soft tissues like muscles or tendons, awkward posture caused by repeated tasks. These are among the most frequently reported causes, resulting in almost 33% of lost or restricted work time

Work Organization and Methods Engineering for Productivity.
DOI: https://doi.org/10.1016/B978-0-12-819956-5.00016-9

2. A well-designed and implemented ergonomics reduces the above strains and plays an important role in performance improvement. This is achieved by designing tasks, work spaces, controls, displays, tools, lighting, and equipment to fit the employee's physical capabilities and limitations, as professed by the Centre for Disease control and Prevention.
3. In addition to reduce the job stresses, ergonomics aims to create safe, comfortable and productive workspaces by designing the work place to suit human abilities and limitations including the individual's body size, strength, skill, speed, sensory abilities (vision, hearing), and even attitudes.
4. By improvement of workplace layout and by reduction of worker strain, engineering ergonomics reduces waste, improves process flow and production metrics, and reduces injury and risk exposure.
5. Manufacturing industries can benefit from ergonomic interventions and innovations, from machine design to process improvement.

16.3 History of ergonomics

The term ergonomics comes from the Greek word, (pronounced as ergon nomos), 'ergon' meaning work, and 'nomos' meaning natural law. It is reported that Greek civilization in the 5th century BC used ergonomic principles in the design of their tools, jobs, and workplaces and the Greek philosopher. Hippocrates gave a description of how a surgeon's workplace should be designed and how the tools he uses should be arranged. Similar evidences are found in other ancient civilizations like Indian, Egyptian and Chinese.

However in the era of scientific management this term Ergonomics was first used by the Polish scientist in his 1857 paper 'The Outline of Ergonomics; i.e. Science of Work, Based on the Truths Taken from the Natural Science'. It entered into English dictionaries after the British psychologist Hywel Murrell used this term at the 1949 meeting of UKs Admiralty, which finally led to the foundation of The Ergonomics Society which is later came to be known as The Chartered Institute of Ergonomics and Human Factors, but subsequently renamed as the Institute of Ergonomics and Human Factors. In USA, the Human Factors and Ergonomics Society (HFES) was founded in 1957. The International Ergonomics Association (IEA) is a federation of ergonomics and human factors societies from around the world.

During the post-world war II era the concept of ergonomic had widely become popular resulting in its becoming the essential requirement for the work study engineers in their projects of redesigning the work places. The Americans replaced the term by 'human factors' (on the lighter side, the Americans prefer to call the lift as an elevator and the flat as an apartment!) emphasizing its applications in non-work-related situations also.

16.4 Definitions on ergonomics

Ergonomics is the application of human biological sciences to achieve optimum mutual adjustment of human and his work the benefits resulting in the increase of efficiency and wellbeing.

I.L.O.

Ergonomics is the scientific discipline concerned with the understanding of interactions among humans and other elements of a system, and the profession that applies theory, principles, data and methods to design in order to optimize human well-being and overall system performance.

International Ergonomics Association.

Ergonomics is an applied science concerned with designing and arranging things people use so that the people and things interact most efficiently and safely, called also biotechnology, human engineering, human factors.

Merriam Webster.

Ergonomics is a science-based discipline that brings together knowledge from other subjects such as anatomy and physiology, psychology, engineering and statistics to ensure that designs complement the strengths and abilities of people and minimize the effects of their limitations.

Chartered Institute of Ergonomics & Human Factors.

Ergonomics is the study of the relationship between workers and their environment, esp the equipment they use.

http://www.dictionary.com.

Ergonomics (also known as human factors engineering), is the science of refining the design of products to optimize them for human use.

https://www.humanscale.com.

Ergonomics is the science of designing the workplace, keeping in mind the capabilities and limitations of the worker.

Ergonomics Plus.

Ergonomics is the process of designing or arranging workplaces, products and systems so that they fit the people who use them.

Dohrmann Consulting.

16.5 Operative's posture at work

All international travelers in flights of duration longer than 8 hours must have experienced the cramps while sitting in one position during the flight.

This is called the static posture of a person which should be avoided. Ergonomics minimizes such unnecessary static work in the workplaces and reduces the forces acting on the body by adhering to the following ergonomic principles:

- All work activities should permit the worker to adopt several different, but equally healthy and safe postures.
- Where muscular force has to be exerted it should be done by the largest appropriate muscle groups available.
- Work activities should be performed with the joints at about midpoint of their range of movement. This applies particularly to the head, trunk, and upper limbs

16.6 The three major domains of ergonomics

The activities of ergonomics can be broadly classified based on their applications, into

- Physical ergonomics,
- Cognitive ergonomics and
- Organizational ergonomics.

Physical ergonomics is based on human anatomy. Some of the anthropometric, physiological and bio mechanical characteristics as related to physical activity are Considered. It analyzes the Work-related musculoskeletal disorders (WRMSDs) and attempts to minimize them.

Cognitive ergonomics is based on mental processes involving human responses like mental workload, decision-making, skilled performance, work stress, etc. which affect interactions among humans and other elements of a system.

Organizational ergonomics is based on the socio-technical systems, involving the organizational structures, Management policies, communications, human relations, teamwork, community ergonomics, cooperative work, new work programs, etc.

16.7 Man-machine system

Man-machine system is a system in which the functions of a human operator (or a group of operators) and a machine are integrated for a coordinated effort between them, that are complementary to each other. This is also commonly referred to as Human-Machine Interaction (HMI), focusing on the design and location of machine controls, by observing the ways in which humans interact with the machine controls as represented by Fig. 16.1.

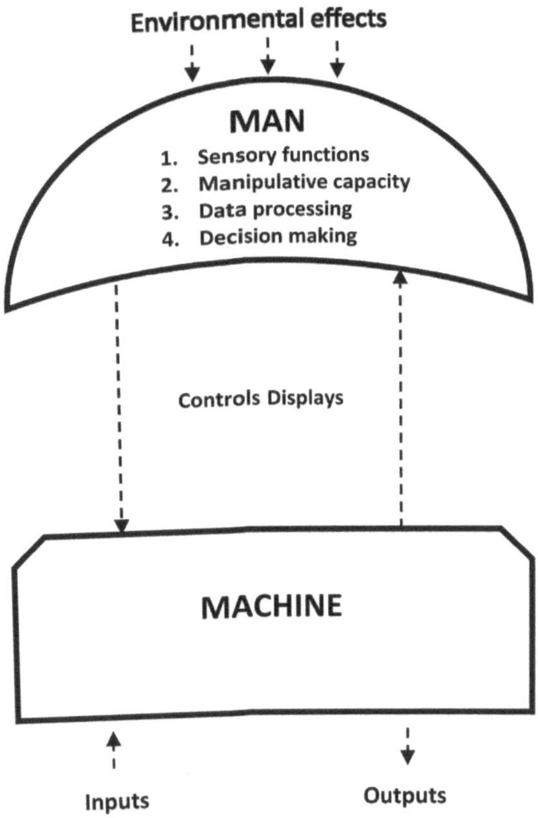

FIGURE 16.1 Man-machine system - human physiology.

The Human-Machine Interaction basically falls into three categories.

Manual systems like assembly operations or using portable drills, etc. Here the operator uses his physical energy with less or no ergonomic aids.

Mechanical systems, like semi automatic machines, where the machine is designed to take the load off the operator. But the location of the displays, use of machine controls and related body movements, call for good ergonomic design of the control and display units.

Automated systems, including today's smart machines, where the machines perform all the operations and controls without human intervention, except for emergency situations. The human intervention is required only in planning, designing, programming and maintaining of the system, which anyway is done elsewhere by computer programming. This system requires low need for operator oriented ergonomic design.

The above concept can also be represented by Fig. 16.2 which shows how the man reacts to the displays and operates the controls.

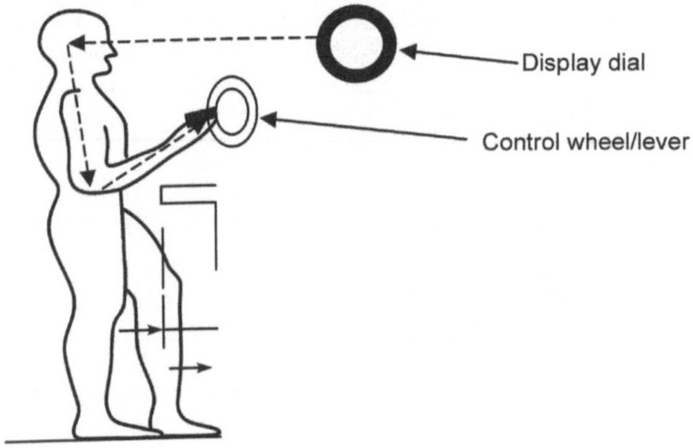

FIGURE 16.2 Man-machine system - how the operator reacts to the displays.

16.8 Ergonomic design of the work place

With regards to the design of workplaces, we can consider four postures of operative's postures.

1. Machine operating posture
2. Small assembly operations while sitting before a work bench, as illustrated in Fig. 16.4
3. Office desk sitting posture
4. Computer operating posture, same as desk work, except that the eye focus or hand stretching is almost constant causing fatigue.

16.9 Ergonomic design of machine controls

This is basically a standing posture with the hands stretched to reach the machine levers and controls. That is why the levers of a machine shall be located nearer to the hands and visual displays should be at the eye level as illustrated in Fig. 16.3 and further detailed below.

- Hand controls like levers, turning wheels should be at the waist level, at a distance not less than 6 and not more than 12 in. from the body position of the operator.
- Push buttons, knobs, toggle switches, etc., should be at the chest level at not less than 6 and not more than 12 in. from the body position of the operator.
- Foot pedals etc. which are used for fast action control with high loading capacity should be located close to the feet location without involving the movement of other parts of the body.
- Visual display units like dials and warning lamps should be at the eye level at not less than 12and not more than 24 in. from the body position of the operator.

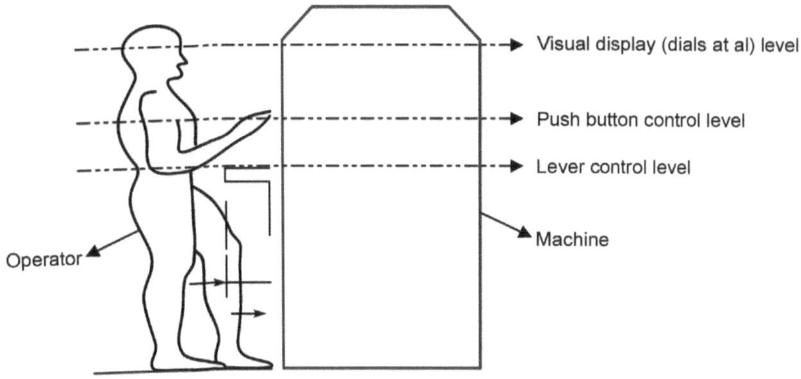

FIGURE 16.3 Man-machine system -location of levers, and controls on the machine.

FIGURE 16.4 Ergonomic design of assembly work place. *From Principles of Wokstudy, ILO.*

16.10 Ergonomic design of assembly work place

Fig. 16.4 from ILOs 1969 book on Work Study illustrates how the assembly work place layout should be organized for simple and logical movement of the hands so as to reduce unnecessary strain on the hand movements.

16.11 Ergonomic design of bins for picking up small components (fig. 16.5)

FIGURE 16.5 The positioning of the bin involving frequent picking up on small parts.

16.12 Ergonomics at office work (fig. 16.6)

The office work table shall be so designed that all desk work materials are located at the normal work area as illustrate in Fig. 16.4, Reference may also be made to paragraph 13.3.2 of Chapter 13 on Principles of Motion Economy, which emphasizes on the ergonomic design of the workplace to achieve maximum motion economy.

16.13 Ergonomics for computer operation

The rapidly changing technology has increased the use of computers in the modern workplace. Unlike the popular concept that Ergonomics is applicable in industrial workplaces only, the increased use of computers and the resulting deskwork has greatly increased the need for ergonomics in the desk work. Though a similar desk as above can be used for computer work, since this involves more of finger movements and eye movements than movement of other parts of the body, the chair must be specially designed for height adjustment, as illustrated in Fig. 16.7.

(i) The computer desk should be ergonomically designed, with provision for CPU and a sliding rack for the keyboard etc. Adjust the position of the computer to prevent Carpal Tunnel Syndrome, as explained in the previous paragraph.

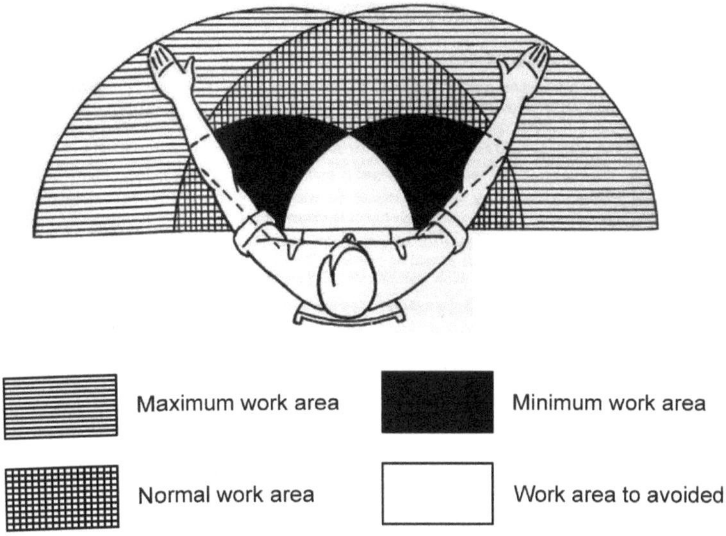

FIGURE 16.6 Ergonomics at office work.

(ii) Even the chairs, monitors and keyboards should be ergonomically designed to make sure that the heights of the desk and chair suit the operative's body stature.
(iii) For computer operators who normally use bifocal spectacles for distance vision, special spectacles with intermediate vision for 3–4 feet to comfortably view the computer screen shall be provided.
(iv) Ensure that the height of the desk chair allows the feet to rest flat on floor
(v) Learn the right way to lift heavy objects to prevent back injuries
(vi) Use handle coatings or special gloves to suppress vibrations from power tool
(vii) The lighting system, the types, design and location of desks, chairs, monitors, keyboards, etc., need to be considered in the workspace, whether it is at the office or at home. The aim is to find the best fit between the operator and his job conditions.

WikiHow puts forward the following 13 principles of ergonomic design of work place for computer operators.

1. Adjust chair height so that your feet are flat on the floor. It is better to adjust the chair height for the correct position of the arms and hands, as most desks are of fixed height. Then, if necessary, provide support for the feet so that the underside of the thighs are just supported at the edge of the seat.

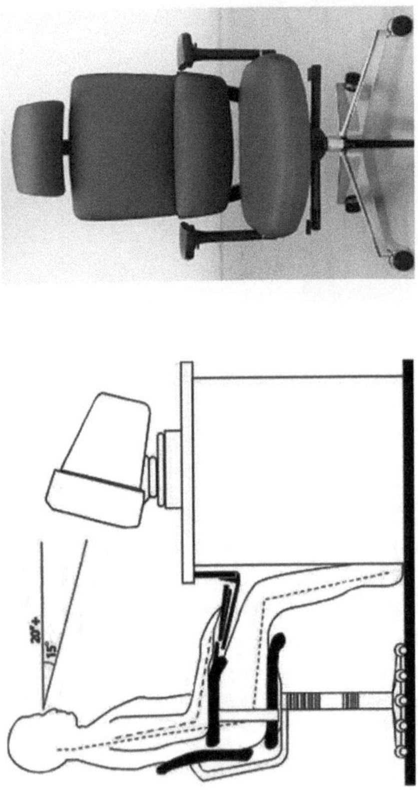

FIGURE 16.7 Ergonomics of computer operation.

2. Keep your wrist position neutral.
3. Change your posture regularly.
    - Sit upright.
    - Sit reclined.
4. Make sure the other adjustments on your chair are correct for you, as well.
5. Adjust the monitor so that the center is at eye level,
6. Adjust the brightness and contrast of your monitor so that you can see clearly and without unnecessary strain.
7. Use a docking station to plug a laptop into a real monitor, keyboard, and mouse.
8. Place your keyboard so that your upper arms hang vertically.
9. Keep your keyboard at the right height.
10. Place the mouse near the keyboard.
11. Manage cables.
12. Arrange items within sight and reach.
13. Get a phone headset.

16.14 Display panels on machinery

Fig. 16.8 illustrates how the controls and the display panels are to be located at ergonomically appropriate levels.

FIGURE 16.8 Location of display panels and control wheels on a machine.

16.15 Management responsibility for optimal ergonomics

Employers are responsible for providing a safe and healthful workplace for their workers. The meticulous application of ergonomic principles would reduce the MSDs due to overexertion of the operatives and subsequently the management is the gainer by way of reduced cost of production.

Especially in high-risk industries like construction, food processing, fire fighting, office jobs, healthcare, transportation and warehousing, the designing of ergonomic process would certainly reduce the risk of developing MSDs. The following are important elements of an ergonomic process:

1. Define the purpose, goals and accountability
2. Provide management support
3. Involve the operative fully in identifying and discussing about hazards in their workplaces and suggesting how to reduce exposure to risk factors. Subsequently the changes made can be assessed form ergonomic point of view. Plan for job rotation for specified jobs, where appropriate.
4. Provide training
5. Identify problems
6. Encourage early reporting of MSD symptoms
7. Develop metrics for accurate and easily understood reporting
8. Maintain documentation and proof of effective application of ergonomic principles.
9. Implement solutions to control hazards.
10. Evaluate progress

The ergonomic training program addresses MSD hazards and should be continuous ongoing process that is incorporated into the daily operations, rather than as an individual project undertaken as a post-mortem.

16.16 Benefits of an optimized ergonomic process

Performance

- Improves productivity
- Improves quality
- Improves process flow
- Improves associate engagement, commitment, and Morale

Cost

- Reduces waste
- Reduces errors
- Reduces & prevents injuries
- Reduces turnover
- Reduces absenteeism and presentism

16.17 Limitations of ergonomics

While most of the design features for ergonomics are as discussed in the previous paragraphs, the actual application and the usage of these locations depends upon the learning capacity on how to use the interface. The other limitations include

1. Ergonomic design discussed earlier may not suit equally to all operatives, in view of different body measurements of operatives, and frequent adjustments may be needed.
2. Take more time and resources increasing the overall costs for short rum operations.
3. Needs very high effort in planning, recruiting, and executing.
4. Requires much longer study periods and therefore requires much goodwill among the participants

16.18 Software for ergonomics

- Ergopoint® from Humantech®, Inc. is an online solution for managing ergonomics
- Medgate's Ergonomics Software - http://www.medgate.com/software/ergonomics/
- Intelex Ergonomics Analysis software - https://www.intelex.com/products/applications/ergonomics-analysis
- JET Ergonomic Assessment Software 'https://ergoweb.com/software/jet/
- 3DSSPP Software from Centre for Ergonomics at the University of Michigan College of Engineering which predicts static strength requirements for tasks such as lifts.
- Ergoweb EnterpriseSM, a New Ergonomic Software System, being released in June 2017, as a service (SaaS) management system for physically demanding work environments.

Benefits of Ergonomic software

The website www.cority.com/ehsqsoftware/ergonomics suggests the following benefits of the above software.

1. Manage organization-wide ergonomic audies and inspections, as well as individual ergonomic assessments.
2. They can effectively reduce ergonomic risks, which could normally be overlooked by us.
3. Provide data base for risk assessment tools, inspection checklists, and self-assessment questionnaires.
4. Consolidate ergonomic findings & actions together in one logical interface.
5. Effectively manage the implementation of ergonomic risk controls.
6. Provide audit reports as per Safety and Occupational Health requirements.

16.19 Conclusion

Understanding or applying Ergonomics is not difficult or complicated, but is mostly guided by common sense. But it is the practical, day-to-day application of these principles that is challenging for many organizations. In a nutshell ergonomics is an essential tool for the sustainability of an organization.

Criteria questions

1. Cite some of the biological sciences that have direct bearing on the operatives at work. (16.1)
2. What is ergonomics and how does it affect the performance of an operative? (16.2)
3. What are the three major domains of Ergonomics? (16.6)
4. Discuss how the control levers should be ergonomically located on the machine. (16.7)
5. What do you understand by Human-Machine Interaction? Explain it's the three variants. (16.7)
6. Illustrate the Ergonomic layout of an assembly work place. (16.10)
7. Illustrate the Ergonomic layout of a computer work table and chair for computer operations. (16.13)
8. It is said that the computer operator should Change the posture regularly. What do you mean by this and why? (16.13)
9. What are the Management Responsibilities for optimal Ergonomics? (16.15)
10. List some of the benefits of an Optimized Ergonomic Process. (16.16)
11. Cite some of the computer Software used for ergonomic layout for the workplace. (16.18)
12. What are the benefits of computer Software used for ergonomics? (16.18)

Further reading

1. International Labour Organisation. *Introduction to work study*. 3rd ed. ILO.
2. Edwards DW. http://www.azquotes.com/quote/873755.
3. Ohno T. http://www.azquotes.com/quote/1411460.
4. Kiran DR. *Production planning and control - a comprehensive approach*. BSP; 2014.
5. Kiran DR, editor. *Proceedings of the work study seminar*. University of Dar Es Salaam; 1983.
6. Kiran DR. *Proceedings of the general management seminar for the transport sector*. Dar Es Salaam: National Institute of Transport; 1984.
7. https://en.wikipedia.org/wiki/Human_factors_and_ergonomics.
8. https://www.osha.gov/SLTC/ergonomics/.
9. http://ergo-plus.com/workplace-ergonomics/.
10. http://www.cority.com/ehsq-software/ergonomics.
11. https://theergonomicexpert.
12. https://piedmontelectric.

Chapter 17

Work sampling

17.1 Principle of work sampling

Work Sampling is a technique to estimate the total time that various activities contribute to the job by taking an assigned number of observations at random intervals. A single observer can collect data on several activities at a time that are impractical or too costly to be measured by time study. This is shorter than production study that involves studying and recording of an operation for 100% of its duration. Work Sampling is hence faster and cheaper. Apart from establishing operational standard times, work sampling can effectively be used to determine production standards, machine and personnel utilization, job allowances, etc.

Originally started to observe the performance of operatives, work sampling has now been used effectively to measure operational times and productivity. It is based upon the laws of probability. A random sample is taken from a large group tends to have the same pattern of distribution as the large group or universe. If the sample is large enough, the characteristics of the sample will not differ much from the characteristics of the group. Typically work sampling can provide information on

1. How much time of the working day during which workers or machines are producing.
2. How much time of the working day used up by delays. The reason for each delay must be recorded.
3. The relative activity of different workers and machines.

17.2 Production study vs. work sampling

Before understanding the principles and uses of work sampling, it is imperative to know something about production study, which involves the recording of all activities and delays that occur on a particular workstation by stopwatch time study right from the beginning of the shift to the end of the shift, or longer if necessary.

As cited in paragraph 17.1, work sampling and production study differ in the duration for which an operation is studied. This aspect can be understood better by comparing the following definitions.

Work Organization and Methods Engineering for Productivity.
DOI: https://doi.org/10.1016/B978-0-12-819956-5.00017-0

17.3 Definitions on production study

Production study is a continuous study of relatively lengthy duration, often extending over a period of one or more shifts, taken with the objective of checking an existing or proposed standard time or obtaining other information affecting the rate of output.

The B S 3138 and also IS 6363.

17.4 Objectives of production study

1. To obtain data to help the management to take a decision on contingency, etc., allowances.
2. To identify the causes for downward trend of the output of a particular workstation.
3. To ensure that the observation and timing adequately covers all work contained in the job and that nothing has been omitted from the work specification.
4. To ensure that each and every activity that occurs during working period is not overlooked, with specific reference to the periodic delays which normally do not occur during the few cycles that are timed by time study.
5. To obtain a detailed account of a performance during a specified period.
6. To validate the standard and allowed times.
7. To assess if the relaxation allowance allowed is appropriate to the job.
8. To identify the avoidable and unavoidable delays as well as other ineffective times.
9. To check levels of output and investigate into unusual or unexpected changes in performance.
10. To establish the reason for different output rates during different times of the day. For example, it is a common notion than the output is higher in the morning than during the afternoon.
11. To validate the synthetic-data, formulae and values in the working situation.
12. As a routine checking, auditing and updating of the existing operational procedure, standard and allowed times etc.

17.5 What is work sampling?

On the other hand, work sampling, which is based upon the laws of probability, the observer keeps visiting a work spot at random intervals and notes down which element or operation is occurring. An end of the day he summarizes the number of the occurrences of individual elements as a proportion of the total number of observations to indicate the overall percentage of each element.

17.6 Definitions on work sampling

Work sampling is a technique in which a large number of observations are made on a group of machines, processes and or workers. Each observation records what is happening at that instant and the percentage of the number of observations recorded for the particular activity or delay is a measure of the percentage of time for which the activity or delay occurs in the total cycle time.

BS 3138 and IS 6363.

Work sampling is the statistical technique for determining the proportion of time spent by workers in various defined categories of activity like setting up a machine, assembling two parts and idle time.

Wikipedia.

Work sampling is a method in which a large number of instantaneous observations are made at random time intervals over a period of time or a group of machines, workers or processes/operations. Each observation records what is happening at that instant and the percentage of observations recorded for a particular activity or delay/idleness is a measure of the percentage of time during which that activity or delay/idleness occurs.

www.yourarticlelibrary.com

Work sampling is a statistical technique used for predicting the total time consumed by an activity, in processes ranging from customer service and manufacturing to telemarketing. It relies on the observations made over a period to record what is happening at that instant, and the frequency at which one or more activities are being performed.

http://www.businessdictionary.com

Work Sampling, also called ratio delay study, is a technique of getting facts about utilization of machines or human beings through a large number of instantaneous observations taken at random time intervals. The ratio of observations of a given activity to the total observations approximates the percentage of time that the process is in that state of activity.

NPTEL, IIT Madras.

Work Sampling is a technique that discovers the proportions of total time that various activities contribute to the job by taking a relatively large number of observations at random intervals.

Auburn University.

17.7 Categories of work sampling

Work Sampling has three main categories:

1. Activity and Delay Sampling
   Activity and Delay Sampling is the most common category, and is used to measure the activities and delays of workers or machines (e.g. to measure the percentage of the day that a person is working or not working).
2. Performance Sampling
   The purpose of performance sampling is to measure working time and non-working time of a person on a manual task, so as to establish a performance index for the person during his or her working time.
3. Work Measurement
   Work sampling can also be used as a work measurement tool to time study several operations simultaneously, to obtain the preliminary time standard so as to select the tasks that should be given priority for time study.

17.8 History of work sampling

- Leonard Henry Tippett of Shirley University, UK, is considered as he father of Work sampling. During his study in British textile mills in 1927 to ascertain the duration and the reasons for the stoppages of the looms, and yarn breakage rates in cloth weaving, He applied statistics and noted that a large number of looms can be studied by making the observations at random. He called this 'snap-reading' method of observation that led to improved production efficiency and operative utilization. He subsequently developed the random numbers table, as well as the Fisher-Tippett distribution.
- R. L. Morrow of New York University, introduced it in US in the name of 'Ratio delay study' since it is basically applied if sample studies in ascertaining the ratio of the delay over the total available time.
- C. L. Brisley and H.L. Waddell later coined the word 'work sampling' in their article published in 1952, in 'Factory management and Maintenance'.

17.9 Why work sampling?

As can be seen from the above points, production study is to be adapted only in specific cases as and when needed. For normal cases of work measurement, especially for operations having shorter cycle times the usual stopwatch time study as explained in the previous paragraph is adapted. By selecting few cycles during the different parts of the day and averaging them serves more or less the same purpose of production study, at the same time enabling us to time more operations in-one day as compared to a single operation possible by production study. In other words, the latter is an example of application work simplification to the work measurement job itself.

Further extension of this work simplification even for the stopwatch time study is work sampling, also called as activity sampling, which involves the estimation of the proportion of time devoted to several of an operation or activity over a certain duration of time by means of intermittent randomly spaced observations. It can tell what percentage of the working day, an operator is actually doing productive work, for how much time he spends for his personal needs or for how long he remains idle due to several specified reasons. For activities of very long duration an appropriate technique of work measurement is work sampling which takes only l/20th of the time required for stopwatch study and gives the result within an accuracy of $\pm 2\%$.

Work sampling is effective in the following two situations.

- For long cycle operations like a maintenance gang working on a group of machines or general foundry operations. Here the emphasis is to establish the causes and time duration of the delays and other ineffective factors that occur during the course of the day than to set up production targets.
- For large groups of identical machines where the operations comprise of similar elements during the course of the day. A Typical illustration is the automatic loom shed in a textile weaving department where one man controls the operation of 12 or more looms. Here the machines work continuously unless they stop automatically due to some defect or other like snapping of the weft or warp yarn. The operator has to overview the working of the machine and when any machine stops, he has to go to the machine, rectify the defect and restart the machine. In this case we are not interested in the %age idle time for one machine, but the average occurrence of stoppage and idleness factor for all identical machines as a group.

17.10 Characteristics of work sampling study

The following factors need to be satisfied for conducting work sampling study:

1. The first and foremost is the non-repetitive work cycles.
2. The second requirement is a long cycle time, during which the observer would be able to record the occurrences of several events.
3. The third is the presence of multiple workers on a production line or assembly line or multiple operations performed by a single worker.
4. There must be enough time available (several weeks or more) to conduct the work sampling study. Since several operations/projects can be observed by work sampling at any time, unlike the time study, the overall time for studying all the operations would still be considerably less.

17.11 Objectives of work sampling

1. To estimate an avoidable delay time & for establishing allowances to standard time.
2. To estimate percentage utilization of machine tools.
3. To estimate percentage of time consumed by various job activities on the part of shop supervisors, engineers, inspectors, repairmen etc.
4. To obtain the results faster and with less expenditure of time and energy
5. To cross check the accuracy of the earlier work standards set by stop-watch time study.

17.12 Procedure for work sampling

The work sampling procedure consists of following three phases

A. *Preparing for work sampling*
   1. Define the problem, determine the main objectives and identify each activity to be measured.
   2. Estimate the measure of the output produced or the types of activities performed on the jobs that are studied, since our ultimate objective is setting work standards.
   3. Obtain the approval of the supervisor of the department in which work sampling is to be performed.
   4. Ensure that all the persons connected with the study; understand the objectives and the procedure adapted by you in conducting the study, and that they accept the frequent visits of the work study engineer.
   5. Select of train the work sampling personnel.
   6. Make a detail plan for taking observations
   7. Determine the time period during which the study will be conducted. Define the starting and stopping points for the study
   8. Decide on the defined accuracy limits for the ultimate results

B. *Performing work sampling*
   1. Conduct a pilot study to
      - estimate the approximate percentage accuracy of activity
      - estimate the required number of observations for the desired accuracy set using the formula illustrated in paragraph 17.15
   2. Based on the random number tables, decide upon the intervals and periods of observations.
   3. Design the actual study and the observation form.
   4. Using this data obtained from pilot study and the estimated value of p, calculate the number of observations to be made.
   5. Observe the activities and record the data. It is better to make the all the observations from the same location
   6. Verify any discrepancy with the supervisor or foreman
   7. Summarize the data at the end of the study.

C. *Evaluating and presenting results of work sampling*
1. Determine the percentages of occurrence of each activity.
2. Analyze the causes for delays which are higher than normal.
3. Propose recommendations and get them implemented.
4. Reconduct work sampling as above for the new method.
5. Compare and evaluate the new and old methods.
6. Present an effective report.

17.13 Statistical theory behind work sampling

Work sampling relies upon the theory of statistical sampling and probability. Theory of normal frequency distribution and confidence level too are associated very much with- work sampling.

The statistical theory of sampling explains that adequate random samples of observations spread over a sufficient period of time can give accurate picture of the actual situations in the system. Approximately 500 observations produce fairly reliable results in most cases, though the actual number of observations to be made depend upon the three factors explained below and incorporated in the formula per paragraph 17.15:

a. The probability of the occurrences of an element. In our case it is the percentage of idle time.
b. The confidence limits with which we want to say that the sample represents the total population or the actual situation.
c. The percentage of error that can be allowed in the study or in other words the accuracy to which you want the result.

Fig. 17.1 below gives a typical, normal curve for which most random observations are expected to conform to from which you can see that the area between x ± 1σ covers 68% of the are under he whole curve or 68% of the samples would fall within these limits. In other words, we can say with

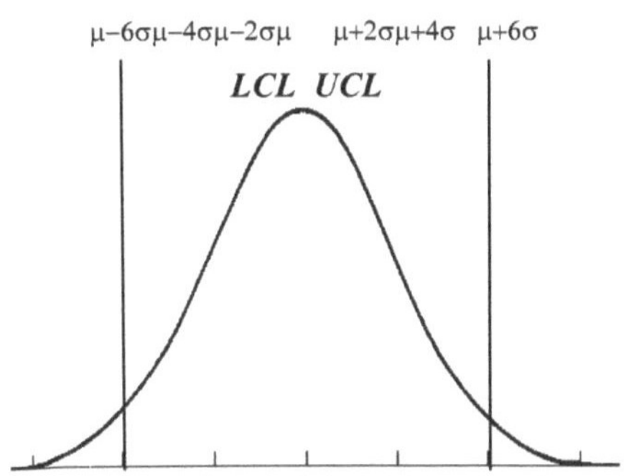

FIGURE 17.1 A typical normal curve and the representation of limits.

68% confidence that the observations between x ± 1σ would conform to presumed situation. This is called the confidence limits.

Table 17.1 below indicates the confidence limits for other values. Similarly, a confidence level of 95.45% signifies that the work study engineer is sure that in 95.45% of the random observation will represent the true factors.

TABLE 17.1 The relationship between value and the confidence limits.

Std. deviation	Confidence interval	Confidence level %	Error level %
0.318 639σ	0.318 639σ	25%	75%
0.674490σ	0.674490σ	50%	50%
0.994458σ	0.994458σ	68%	32%
1σ	1σ	68.2689492%	31.7310508%
1.281552σ	1.281552σ	80%	20%
1.644854σ	1.644854σ	90%	10%
1.959964σ	1.959964σ	95%	5%
2σ	2σ	95.4499736%	4.5500264%
2.575829σ	2.575829σ	99%	1%
3σ	3σ	99.7300204%	0.2699796%
3.290527σ	3.290527σ	99.9%	0.1%
3.890592σ	3.890592σ	99.99%	0.01%
4σ	4σ	99.993666%	0.006334%
4.417173σ	4.417173σ	99.999%	0.001%
4.5σ	*4.5σ*	*99.9993204653751%*	*0.0006795346249%*
4.891638σ	4.891638σ	99.9999%	0.0001%
5σ	5σ	99.9999426697%	0.0000573303%
5.326724σ	5.326724σ	99.99999%	0.00001%
5.730729σ	5.730729σ	99.999999%	0.000001%
6σ	6σ	99.9999998027%	0.0000001973%
6.109410σ	6.109410σ	99.9999999%	0.0000001%
6.466951σ	6.466951σ	99.99999999%	0.00000001%
6.806502σ	6.806502σ	99.999999999%	0.000000001%
7σ	7σ	99.9999999997440%	0.000000000256%

It may be noted that this σ value is different in concept from sigma level (like six sigma level) as used in TQM Chapter 24 of the book, Total Quality Management, Key Concepts and Case Studies, referred under Further Reading for better clarification.

Let us presume that the work study engineer takes 25 rounds of the machine shop in a day to observe the operator 'x' and finds that.

- 15 times he is working on the machine
- 4 times he is setting tool or cleaning the machine
- 3 times he is not doing anything
- 3 times he has gone for his personal needs.

This shows that the worker spends 60% of his time in actually working over machine and for 12% of the total time he is idle, etc. These facts can be confirmed by conducting more number of observations.

17.14 Random timing

We have seen, in the earlier paragraphs that random sampling allows every unit or part of the total population on equal chance of being included in the sample. That is the parameters of the sample chosen at random are expected to be same as those of the whole population. In the work sampling, the randomness is in the timing of observations, so that the readings taken at these intervals represent the actual situation. To avoid individual bias, a pre-prepared list of random tables can be referred to. Nevertheless in these days of computerized data processing you can prepare your own random tables. The following procedure is suggested for the use of these random tables depending upon the number of observations you want to make and the time available for you to complete all the observations.

1. Choose the numbers in order they appear in the table or every alternative number or every fifth number etc., depending upon the total observations you want to make, say 22, 60, 75, 2, 90, 48, 36, 42, 99, 55, 10, 17, 52, 88, 26.
2. Arrange these numbers in ascending order, i.e. 2, 10, 17, 22, 26, 36^48, 52, 55, 60, 75, 82, 88, 90, 99.
3. Depending upon the total time available with you before which you have to complete the observations and the range of these random numbers assign a time value for the unit. For example, if the number range between 0 to 100, and you have to complete talking 15 observation in a shift of 400 minutes assign a value of 4 minutes for each unit. That is the observations will be at 8, 40, 68, 88, 104, 144, 192, 208, 220, 240, 300, 328, 352, 360 and 396 minutes after the start of the study say 8:00 a.m.

By using the random tables as above, the personal bias in choosing the time of each visit can be eliminated. Also, if the worker knows the definite time of his next visit, he might deliberately feign productive time at that time, pushing all unproductive elements to the non-visit periods. This will naturally give distorted results.

17.15 Number of observations to be made

From statistics, the accuracy of the study is given by

From this we can calculate n, the number of observations as

$$n = \frac{pq}{\sigma_p^2} \text{ or } \frac{p(1-p)}{\sigma_p^2}$$

17.16 Use of nomographs for determining sample size

Other methods for finding one of observations are by use of nomographs, also called nomograms or alignment charts.

A nomograph is defined as a graph, usually containing three parallel scales graduated for different variables so that when a straight line connects values of any two, the related value may be read directly from the third vertical line at the point intersected by the line.

For example, if the percentage of occurrence during the initial study is 5.5% and the accuracy needed is 98.1%, then the no. of observations to be made are 620 at 95% confidence limits, as illustrated in Fig. 17.2.

17.17 Advantages of work sampling

(a) It involves much less cost compared to stopwatch time study.
(b) Does not require the analyst to continually observe the job.
(c) It can be performed with much less training.
(d) It can be used for long operations, which are almost impractical to be measured by stop watch time study.
(e) Very advantageous for timing group activities.
(f) It doesn't need any timing device like stopwatch or micrometer, etc.
(g) Even if the study gets interrupted in between, it does not introduce any error in the results.
(h) Observations can be made within the desired accuracy.
(i) Large number of observations extended over days or weeks reduces the influence of day to day fluctuations on the results.
(j) It can increase efficiency by uncovering the sources of delay.

17.18 Limitations of work sampling

1. It is uneconomical and less accurate, both as regards time and money to study activities of short duration by work sampling.

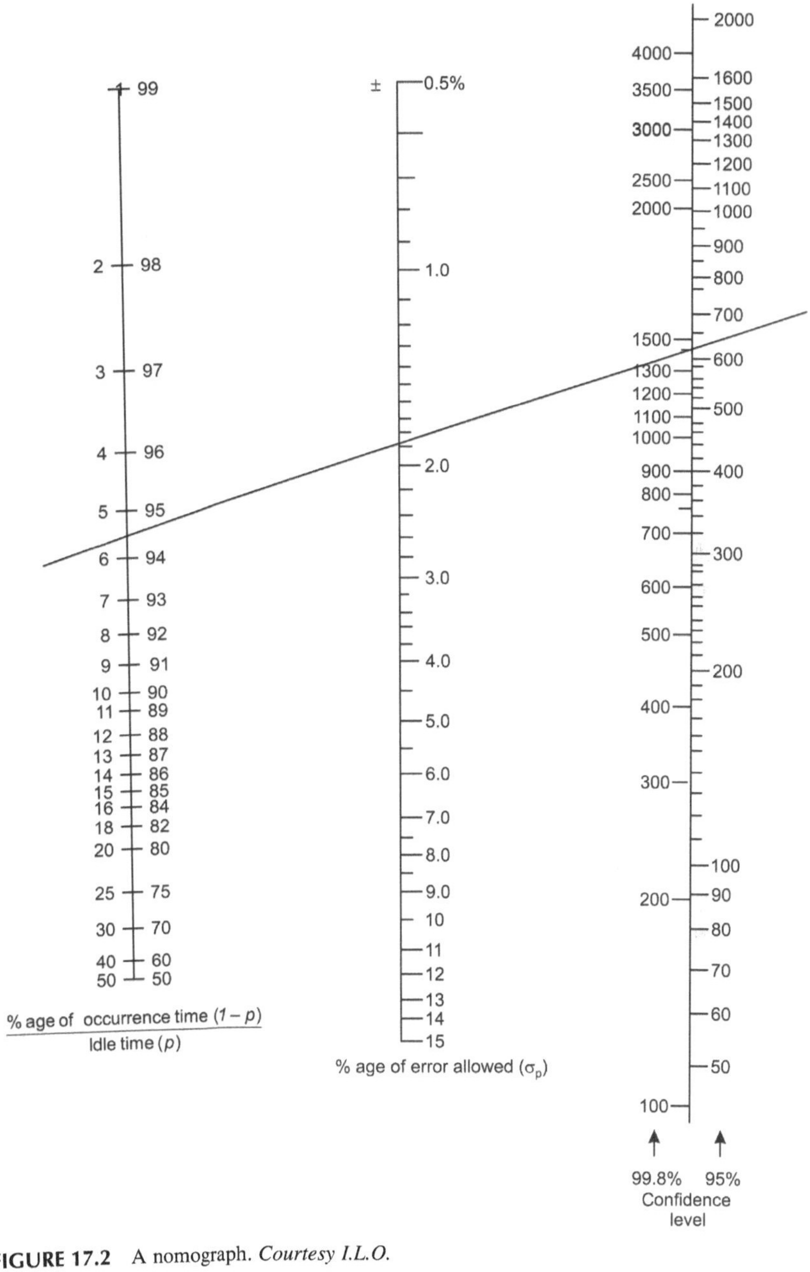

FIGURE 17.2 A nomograph. *Courtesy I.L.O.*

2. It is also uneconomical if only one worker or one machine is to be studied.
3. It does not break the job into elements and hence does not provide element details. as completely as time study.
4. It does not assist in the improving of work methods.
5. It normally does not account for the speed at which an operator is working.
6. Workers may not understand the principles of work sampling and hence may not trust it.
7. Observations, neither random nor sufficient in number may produce inaccurate results.
8. Can yield biased results if observer does not follow random pattern.

17.19 Applications of work sampling

To summaries, we can cite the applications of work sampling as under

1. To determine the working and idle times of men and machines.
2. To time long duration activities which are regular/irregular or frequent/infrequent.
3. To identify unwanted material handling don/ in day.
4. To estimate allowances for unavoidable delay.
5. To describe resources utilizing pattern.
6. For the purpose of cost control accounting.
7. To estimate the percentage utility of inspectors and time standards for indirect labor cost.
8. Very useful for stores, hospitals, ware housing, offices, farm work, repair and maintenance work in textile mills, machine shop, etc.
9. It is preferred when the cost of using other work measurement technique of timing a job appears to be great.

17.20 Performance sampling

This is a work sampling process, wherein the observer indicates the rating of the operative at the time of the observation on a suitably designed format and evaluates the readings using this rating factor. This will present a more accurate picture of the percentages of occurrences.

17.21 Computer software for work sampling

There are several software available for data entry and analysis of the elements performed during the sampling rounds. Most of these can also be loaded on to smart phones for facilitating on the field studies. Some popular software are cited below.

1. **WorkStudy +**

   WorkStudy + is ideal for any type of data sampling, including fixed and random interval sampling. With a built-in sampling timer. WorkStudy + will automatically alert users when it is time to collect the next sample.

2. **PATHSoft**

   PATHSoft was developed by South Eastern Louisiana University, PATH implying Posture, Activities, Tools, and Handling methodology. In addition to the web-based portion, this software has a Windows Mobile local application for use with Personal Digital Assistants (PDAs) and other mobile devices.
3. **laubrass' UMT Plus software**

   This software possesses random study and multi-study features of UMT Plus to create performance standards for traditional time-work positions.
4. **The Work Sampling System**

   The Work Sampling System is a curriculum-embedded, authentic performance assessment for teachers to use to assess young students' development in multiple domains. The program features updated content, based on Common Core and state early learning guidelines and current research in the field of early childhood education.

17.22 Conclusion

As stated in the synopsis, Work sampling is a process, where many instantaneous observations made at random time intervals over a period of time would yield fairly accurate time standards which can substitute the laborious process of work measurement. This is based on the theory of probability. Hence it is essential for us to understand how the theory of probability would lead to the success of work sampling, as explained in this chapter.

Criteria questions

1. Distinguish between Work Measurements. (17.2)
2. What are the Objectives of Production Study? (17.4)
3. Distinguish between Activity and Delay Sampling and Performance Sampling. (17.7)
4. Trace the History and Development of Work Sampling. (17.8)
5. Enumerate the Requirements of Work Sampling. (17.10)
6. Discuss the Procedure for conducting Work Sampling. (17.13)
7. Why is it important to make sure that Workers have understood the Purpose of, and accept the Frequent Visits? (17.13)
8. Why is it necessary to Time the Visits at Predetermined Random intervals and not at Regular Periods? (17.15)
9. How do you determine the Number of observations to be made for a Particular Operation? (17.16)
10. What are nomographs? How they are useful in Work Sampling? (17.17)
11. What are the Limitations of Work Sampling? (17.19)
12. Cite some of the Computer Software available for Work Sampling. (17.22)

Further reading

1. Barnes R. *Motion and time study*. 7th ed. 1980.
2. Kiran DR. *Total quality management, an integrated approach*. BSP; 1963.
3. Kiran DR. *Production planning and control, a comprehensive approah*. BS Publications; 2018.
4. Brisley CL. Work sampling. In: Maynard. editor. *Industrial engineering handbook*. 3rd ed. McGraw.
5. Groover MP. *Work systems and methods, measurement, and management of work*. Pearson Education International; 2007.
6. Sheth V. *Industrial engineering methods and practices*. Penram International Publishing; 2000.
7. Browker A, Liebermann G. Industrial statistics. In: Ireson WG, Grant EL, editors. *Handbook of industrial engineering and management*. Prentice Hall; 1971.
8. Tsai W-H. A technical note on using work sampling to estimate the effort on activities under activity-based costing. *Int J Prod Econ* 1996.
9. http://www.southeastern.edu/acad_research/depts/iet/news/archive/drs_yuan_and_alkadi.html.
10. www.laubrass.com/umtplus.
11. www.umtproducts.com/casestudies/work-measurement-reviews/retail-productivity.

Chapter 18

Value analysis

18.1 What is value analysis?

Value Analysis is a major cost reduction and control techniques with systematic approach to ensure that a specific product or component is designed and manufactured to serve all the desired functions at the minimum cost without diminishing quality, reliability, performance and appearance. Since the functional value of a component is analyzed with reference to the design and engineering aspect, this technique also is called value engineering. It is a systematic and creative analysis to identify the functions of a product or a component and to provide the desired function at the lowest total cost. By eliminating the unnecessary design features which add neither to quality nor to the appearance of the product.

In short value analysis involves critical examination and analysis of the design of a component akin to the modern DFSS tool of DMAIC, with reference to its functional value. It is a systematic and creative method using proven method, to obtain the same or better performance at a lower cost so as to improve the value of goods or products and services by critical examination of its function. Value Engineering seeks to optimize value for the money in projects and emphasizes on analyzing the functional values of all design features of a component making it a special among all DFSS tools.

It has been used in almost every kind of application. It helps people creatively generate alternatives to secure essential functions at the greatest worth referred to as value as opposed to costs. It is also known as Value Management, Value Planning, Value-Added Value Engineering, Functional Analysis and a host of other names as illustrated in paragraph 18.3.

18.2 Definitions of value analysis

Some definitions of Value Analysis, which also give an insight into the technique, are:

A discipline action system, attuned to one specific need: accomplishing the functions that the customer needs and wants at the lowest cost.

Miles (1972).

Work Organization and Methods Engineering for Productivity.
DOI: https://doi.org/10.1016/B978-0-12-819956-5.00018-2

A proven management technique using a systematized approach to seek out the best functional balance between the cost, reliability and performance of a product or project.

Zimmerman (1982).

Systematic analysis that identifies and selects the best value added alternatives for designs, materials, processes, and systems. It proceeds by repeatedly asking "can the cost of this item or step be reduced or eliminated, without diminishing the effectiveness, required quality, or customer satisfaction?"

Business Dictionary.

Value Analysis/Value Engineering as an intensive, interdisciplinary problem solving activity that focuses on improving the value of the functions that are required to accomplish the goal, or objective of any product, process, service, or organization.

Value Engineering: The Forgotten Lean Technique by James R. Wixson.

Value engineering is a systematic and organized approach to provide the necessary functions in a project at the lowest cost. Value engineering promotes the substitution of materials and methods with less expensive alternatives, without sacrificing functionality. It is focused solely on the functions of various components and materials, rather than their physical attributes.

Investopedia.

Value engineering can be defined as an organized effort directed at analyzing features, systems, equipment and material selection for the purpose of achieving the essential functions at the lowest life cycle cost consistent with the required performance, quality, reliability and safety.

US General Services Administration.

The value methodology (VM) is a systematic and structured approach for improving projects, products, and VM, which is also known as value engineering, is used to analyze and improve manufacturing products and processes, design and construction projects, and business and administrative processes.

SAVE.

A professionally applied, function oriented, systematic team approach used to analyze and improve value in a product, facility design, system or service.

SJVE.

18.3 History of value analysis

Value Analysis was conceived by Lawrence Miles at General Electric Co. in 1945 based on the application of function analysis to the component parts of a production shop in view of shortages of skilled labor, raw materials, and component parts during World War II. While America adapted the term Value Engineering, Europe adapted the term Value Analysis. By 1952 value engineering began its growth throughout industry.

The Public Law 104–106 of US states, "Each executive agency shall establish and maintain cost-effective value engineering procedures and processes."

The Federal-Aid Highway Act of 1970 too has recognized the effectiveness of value analysis, made the first Federal Highway reference to VE, requiring that "in such cases that the Secretary determines advisable plans, specifications, and estimates for proposed projects on any Federal-Aid system shall be accompanied by a value engineering or other cost reduction analysis."

Some of the acronyms related to value Engineering are

- VA/VE Value Analysis and Value engineering
- VM Value Management
- VMd Value Methodology
- VAVE Value Analysis/Value Engineering
- FA Functional Analysis
- FAST Functional Analysis System Technique
- SAVE Society of American Value Engineers
- SJVE Society of Japanese Value Engineering
- INVEST Indian Value Engineering Society
- INVAVE Indian Journal of Value Analysis/Value Engineering.
- DARSIRI Data (collect), Analyze, Record ideas, Speculate, Innovate, Review and Implement
- CSVA Canadian Society of Value Analysis
- IVM UK Institute of Value Management

18.4 What is value?

Here the term value is distinguished from price or cost because value is more an abstract concept referring to the cost benefit aspect. It is the ratio between a function for customer satisfaction and the cost of that function. Functional worth is the lowest cost to provide a given function.

Value is a perception, hence customers will have their own perceptions on how they define value. However, we can relate value to quality, performance, style, and design in comparison to the product cost.

In fact, value is the performance of a product in relation to the function what the product or service is supposed to do in relation to the cost and expenditure needed to create it.

Value =

If the numerator is increased without increasing the denominator, Value increases. In other words, a product can be engineered for improved value by either increasing the quality, reliability, availability, maintainability, serviceability etc., for the same cost or by reducing cost for the same degree of the above factors of quality, reliability etc. In short value analysis results in more functions at the same cost or same function performed at lesser cost. The manner in which the thinking is applied gives the concept its name value analysis, and the planning the analysis in the design function gives it the name value engineering. We can say that Value Engineering is the functional aspect of Value Analysis, which refers to the analytical overall concept. This chapter hence uses the term value analysis as a synonym to the term value engineering, as cited in paragraph 18.3.

18.5 Value analysis

As stated before, value analysis is a methodology to increase the value of an object. Value analysis is a planned, scientific approach to cost reduction, which reviews the material composition of a product and production design so that modifications and improvements can be made which do not reduce the value of the product to the customer or to the user. The object to be analyzed could be an existing or a new product or process, and it is usually accomplished by a team following a work plan.

18.6 Objectives of value analysis

1. To reduce production and total cost as illustrated by the case study per paragraph 18.12.
2. To improve operational performance.
3. To improve product quality.
4. To reduce the manufacturing costs.
5. To improve customer-supplier relations.
6. Cost avoidance on future programs.
7. To reduce in product variations.

18.7 Typical benefits of value analysis projects

The benefits of value analysis, which fulfill the above objectives, can be summarized as below:

- Value analysis aims to simplify products and processes. Thereby increasing efficiency in managing projects, resolve problems, encourage innovation and improve communication across the organization.
- Value analysis enables people to contribute in the value addition process by continuous focus on product design and services.

- Value analysis provides a structure through cost saving initiatives, risk reduction and continuous improvement.

18.8 Functions of a product as the customer sees

Value Analysis is based on a study of functions of a product or service. It involves the identification of functions from the knowledge of the customer needs. The first approach to the identification of functions should be focused on basic functions. These functions are those for which the customers believe they are paying. There are usually only one or two basic functions per product or service.

All the functions can be grouped as below as per their levels of importance:

- The basic function which is the very purpose of the product or service.
- The secondary functions are those not directly accomplishing the primary purpose, but support it from a specific design approach. These can also be subcategorized as use functions or aesthetic functions.
- Use functions are those, which answer the question how the basic function is achieved. For example, if the primary purpose of a bottle is to contain a liquid, the secondary purpose can be strength to support the contents even when dropped or transparency so that the contents can be identified without opening the bottle.
- Aesthetic functions, whose purpose is only to add beauty or esteem value to the product and are associated with feelings. In the above example, the attractive color or shape provides the aesthetic function.

It is generally found that the primary functions are achieved by 20% of the total cost whereas the secondary functions account for 80% of the cost. This is the crux of Value Analysis.

Once you identify the functions, they must be written down in 2 words, a verb and a noun, as further explained in paragraph 18.10.3.

18.9 Functional value of a product

Functional value or use value of a product is the purpose the product fulfills and is an attribute that provides the customer with functional utility, which can be distinguished from other values of a product as illustrated below.

- **Cost value:** is the cost of manufacturing and selling an item
- **Exchange value:** is the price a customer is prepared to pay for the product, or service
- **Place value:** Same item may have different values at different places. For example, a glass of water in a desert.
- **Time value:** An item may have a high value at certain point of time. Once the time is passed, it may lose its value. For example, blood transfusion to a patient during an operation.

- **Esteem value:** is the prestige a customer attaches to the product
- **Fancy value:** Apart from the prestige or esteem value of possessing an esteemed commodity, this author wishes to introduce a new type of value called fancy value, when buyers consider the value of a commodity to be directly proportional to the selling price and buy them just for the fancy of purchasing a costly item. For example if a certain pair of shoes of an certain brand and quality are sold for a certain price at one shop, and an identical pair of shoes of the same brand and quality are sold at 10% higher price in a mall, some people prefer to buy from the higher priced shop, just to satisfy their fancy of buying a commodity in a fancy mall

18.10 Methodology of value analysis

The creative mind required for value analysis makes it comparable to the method study that analyzes and improves the manufacturing operations, while value engineering analyzes and improves the design factors. We can say

Method study: production:: value analysis: design

Hence the methodology for value analysis is similar to the SREDDIM of Method study, as explained in Chapter 6 on Method study. Nevertheless, in case of value analysis, it is split into 8 phases, the terminology for each phase being as follows:

The 8 phases of value analysis

1. General phase
2. Information phase
3. Functional phase
4. Investigation phase
5. Creative phase
6. Evaluation phase
7. Recommendation phase
8. Follow-up phase

18.10.1 General phase

After identifying the existing product or the process to be analyzed, its general description is given indicating the functions and design features etc., of the product as well as its components. List the basic functions (the features for which the customer is paying), as identified by the function phase.

18.10.2 Information phase

Additional data like the operational sequences or the time standards are recorded. These data would assist in analysis and in the comparison of the proposed process with the existing process.

18.10.3 Function phase

Identify and list all the functions of the product or process, for which the customer is paying. Here it is necessary to indicate each function in only two words, a noun and a verb. This enables conciseness. By trying to describe a function in a sentence, we may unwittingly combine 2 or more functions which would cause confusion in our analysis. Table 18.2 of the case study provides an illustration to this concept.

Again, while identifying a function, specify it so as not to limit the ways in which it can be performed. For example, don't say 'screw nameplate' but say 'attach a nameplate', since the nameplate can be attached not only by screwing but also by soldiering, riveting, or gluing etc. The later specification would help us in thinking of alternative solutions for this function.

Once all the functions are listed, isolate the basic function followed by all the secondary functions. This will help in our analyzing each of the secondary function is really necessary or can be done away with. Given below are the criteria and guidelines, as already illustrated in paragraph 18.8, to distinguish between the basic and secondary functions

- Basic function is the primary reason for an item or system. It is the performance feature that must be achieved if it has to perform its purpose.
- A secondary function is the features of an item which supports the basic function, and even without that function, the item can perform its functions. For example, the primary function of paint is to protect the surface, while the secondary function is to give a good appearance.

Guidelines for defining the functions

- The function shall be defined only by 2 words, a verb and a noun.
- The noun shall be measurable and/or countable.
- The noun shall as far as possible signify the design based constraint.
- The verb shall be active and effect the noun directly.
- The function shall be verifiable.

18.10.4 Investigation and creative phases

While all required data is collected and recorded in the investigation phase. It is the creative phase and is the heart of the methodology. Since these 2 phases overlap each other, they are discussed together in this paragraph. All the points discussed in paragraphs 8.5–8.7 including the brainstorming, in the chapter on 'Examine and Develop' with respect to creativity are applicable here.

The objective of this phase is to find a better way to do the main function by asking the following questions for each of the identified function and determine the relative importance of each function, preferably by asking a representative sample of customers.

- Does it contribute value? (Is there something that does not contribute value?)
- Is the cost in proportion to the function realized?
- Does it need all its parts, elements, procedures?
- Is there something else to do the same function?
- Is there a standard part that can do this function?

18.10.5 Evaluation phase

- Each idea generated should be analyzed and developed in a manner to be more logical and practical making it function better.
- Identify barriers like mind set concepts opposing the idea and discuss whether the barriers hold strongly against the ideas. Isolate and eliminate them, but after recording them for future reference.
- Choose two to four ideas among them and make a comparative study regarding the cost as well as performance.

18.10.6 Recommendation and follow-up phases

After all, since any analytical study has to be approved by the top management, it is hence imperative that the value analyst team prepares a report detailing the several factors considered as detailed earlier emphasizing the net cost saving as well as the functional improvements achieved and submit the same to the top management as their recommendation. Once the recommendation is accepted, the operatives and other related personnel will have to be trained and regular follow-up with the implementation has to be maintained. This phase is similar to the steps Install and Maintain of method study.

18.10.7 Darsiri methodology for value analysis

Some books cite a 7 step DARSIRI methodology which is similar to the above 8 phases and as follows,

1. Data (collect),
2. Analyze,
3. Record ideas,
4. Speculate,
5. Innovate,
6. Review and
7. Implement.

18.11 Function analysis system technique (FAST)

Function Analysis System Technique (FAST) is a graphical representation of the functions identified during the value analysis program. It builds upon value analysis by linking the simply expressed, verb-noun functions to describe complex systems in a logical sequence, visualize the need for and role of each major component and prioritize them.

18.12 Case study

The JLO division of the Surat Unit of the Ralli group manufactures a 25 cc petrol engines used for agricultural sprayers and cycle rickshaws. A major component of this engine is the clutch base plate of the transmission assembly, whose main function is to absorb the thrust exerted by the clutch plate during the power transmission. This being the costliest component of the engine, a value engineering study was conducted as detailed below. This would provide a clear understanding of the methodology and benefits of value engineering.

Component: Clutch Base Plate in the Transmission Assembly of a Petrol Engine

(i) **General phase:** The general information about the function of the Clutch Base Plate is as follows.

The 25 cc JLO petrol engine, used in agricultural sprayers and cycle rickshaws has a clutch base plate as a part of the power transmission system. This component is under study being one of the costliest components in the transmission assembly (Fig. 18.1), as described briefly as under.

The drive from the clutch shaft is through the gear (which rotates freely on the clutch base) and the friction plates on one side and the clutch plate, the clutch base and the clutch shaft on the other side. In the normal running, the friction plates hold the gears tightly against the clutch plate, (and hence on to the clutch base), transmitting motion by friction under pressure. When the clutch pin is released the friction plates move to the right against the spring pressure, thus releasing the pressure between the gear and the clutch plate, thereby enabling the gear to rotate freely on the clutch base.

(ii) **Information phase:** The clutch base is machined from 80 mm dia EN8 steel bar, as per Fig. 18.2. The sequence of operations (Table 18.1) is as under.

(iii) **Functional phase:** The functions of the clutch base plate and their functional levels are indicated in Table 18.2.

It may be noted that the primary function is specified as 'Absorb Thrust' while others like facilitate Drive (from pinion to drive shaft) are categorized as secondary. This conforms to the fact that by fixing

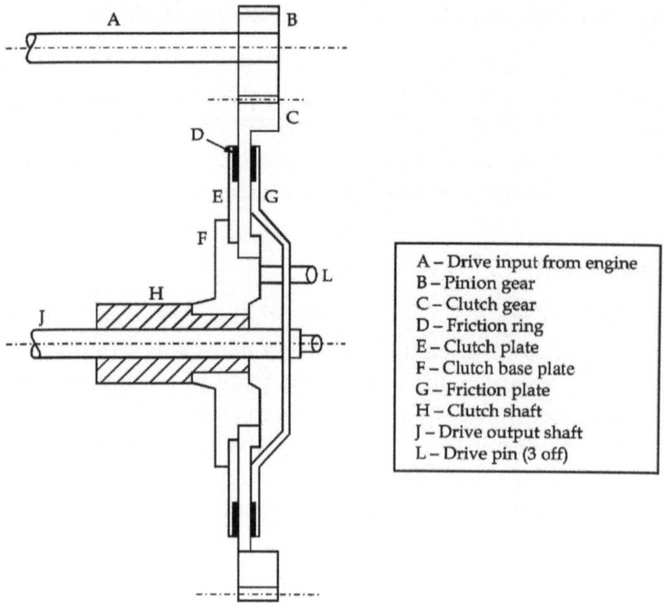

FIGURE 18.1 Clutch transmission assembly.

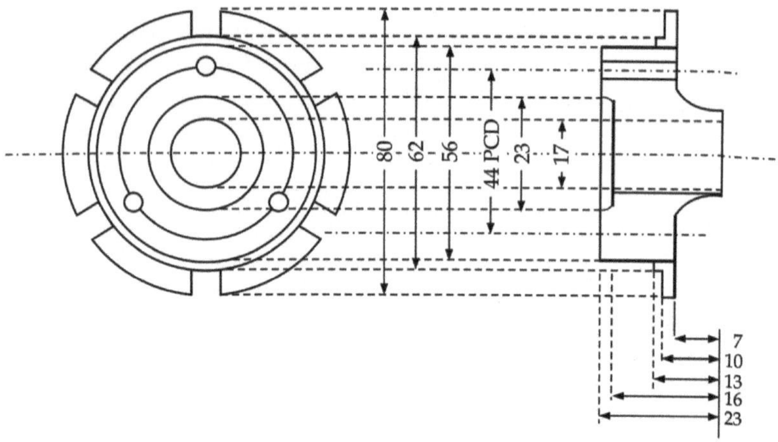

FIGURE 18.2 Clutch base plate.

the clutch plate firmly onto the drive shaft and making the gear rotate freely on it, it may be possible to transmit the drive through the clutching and declutching action. But the spring thrust is so high that the clutch plate would fail and it is the clutch base plate that will absorb the thrust and thereby enable the friction and the clutch plates to perform the clutching and declutching action and transmit the power without deforming the plates.

TABLE 18.1 Operation sequence.

Machined on bar stock			Machined on forging		
Op. No.	Operation	Std. time (m)	Op. No.	Operation	Std. time (m)
1	Cut blanks 28 mm thick	2.60		–	
2.	Load on lathe, face one side, drill 17 dia and counter bore	8.00	1	Load on lathe, face one side, drill, and ream 17 dia and counter bore	8.20
3	Bore 17 mm dia	5.80		–	
4	Fix other side on chuck, face and turn 23 mm dia	8.00		–	
5	Countersink	0.50	2	Countersink	0.50
6	Copy turn first side	4.20		–	
7	Copy turn second side	3.93		–	
8	Broach key way	2.50	3	Broach key way	2.50
9	De-burr key way	1.00	4	De-burr key way	1.00
10	Mill 6 slots	6.75	5	Mill 6 slots	6.75
11	Drill 3 holes	1.50	6	Drill 3 holes	1.50
12	Countersink 3 holes	0.80	7	Countersink 3 holes	0.80
13	Tap 3 holes	1.50	8	Tap 3 holes	1.50
14	Grind 55.3 dia	3.00	9	Grind 55.3 dia	3.00
15	Grind 62 mm dia	2.00	10	Grind 62 mm dia	2.00
16	Face the boss	1.00	11	Face the boss	1.00
17	De-burr, coat anti rust and store	3.00	12	De-burr, coat anti rust and store	3.00
	Total time	**56.08**			**31.75**

(iv) **Investigation phase:** If we analyze the costs involved in sustaining each of the 5 functions, we can notice that the total cost of maintaining the secondary functions account to as much as 80% of the total cost. Especially the secondary function of 'facilitate drive' can be rated as the costliest due to the cost of providing the flange and the slots.

This should give an idea as to which design features should be questioned more thoroughly to get maximum cost reduction by changing only the minor and secondary design features.

(v) **Creative and evaluation phases:** While the finished piece of the clutch base plate is 300 gm the raw material used is 80 mm blank weighing 1200 gm, thus 75% of the material being lost as scrap.

Table 18.1 pin points the operations that take unusually high machining time. The first step in the application of creativity is to identify the alternative processes to replace the high cost operations viz 2 and 4 (facing), 3 (boring) and 10 (milling) which account for 90% of the total operational time.

Hence the flange feature of the existing design, which contributes to abnormally high cost, is considered for value analysis and 3 alterative designs are proposed.

(vi) **Recommendation phase:**

(a) **Design change No. 1:** Changing the RM specification to forged steel. Change the raw material to forged steel instead of a cutting from an 80 mm EN8 (BS970) steel bar. The clutch base shall be made from forged steel with a maximum of 1 mm excess material where only at these points where machining is required, This will reduce the machining time to a bare minimum besides reducing the scrap to 15%. The total machining time is estimated at 8 minutes.

- The quoted weight of the forged blank of specification EN 8 steel is 500 gms and the price of this forged piece is Rs. 7.25 per piece (Rs. 6.25 for IS 226-MS).
- Table 18.3 below compares the costs between fully machined and forged component, yielding a direct saving of Rs 4.55 per piece by replacing bar stock with forged pieces.

(b) **Supplementary design change No. 2:** Integration of the flange portion of the base plate with the clutch plate itself by screwing (Fig. 18.3).

By looking at the component and its assembly with the clutch plate, it can be deduced that the flange portion and its milling is designed purely to hold the clutch plate and to receive positive power transmission. The axial movement of the plate with respect to the base is not required. In fact, there were initial complaints that the plate tends to slip into the gap between the gear and base when the clutch is released.

Analyzing the available bar sizes and the ideal overlapping, a bar size of 75 mm dia is considered optimal with a potential saving is Rs 30,000 per annum.

TABLE 18.2 Functions of the clutch base plate.

Sl. No	Functions		Functional level		Remarks
	Verb	Noun	Primary	Secondary	
1	Facilitate	Drive	–	Yes	To pinion
2	Resist	Bending moment	–	Yes	
3	Support	Weight	–	Yes	Of the reduction assembly
4	Absorb	Thrust	Yes	–	From the clutch plate
5	Provide	Location	–	Yes	For drive pin

TABLE 18.3 Cost comparison in changing to Forged steel.

Sl. No	Cost element	Fully machined component		Forged component	
		Details	Cost in Rs.	Details	Cost in Rs.
1	Raw material cost	1.2 kg of EN-8 bar at Rs 5.75/kg	6.90	Vide quotation	7.25
2	Hacksaw cutting	Sub-contracting	2.00	Nil	–
3	Machining	**56.08** m	7.10	31.75 m	4.20
	Total		**16.00**		**11.45**

Note: The prices and costs were as prevalent in the 1970 when this study was undertaken. Nevertheless, the comparison between the existing and the proposed processes is valid as on date. Fig. 18.2 indicates the dimensions of the clutch base plate after machining. It is to be noted that there is no dimensional variations of the machined piece, whether from bar stock or from forged stock.

Salient features of this design change are

1. The bar stock size is reduced from 80 mm to 75 mm resulting a further saving of Rs 3.45 core piece in material cost and Rs 1.50 per piece in machining cost, i.e. Rs. 4.95 per piece. The total saving by implementing both the recommendations is

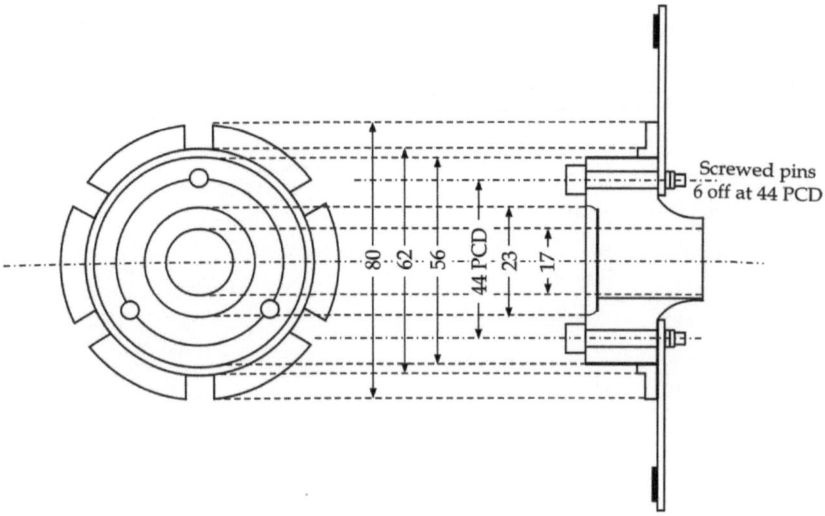

FIGURE 18.3 Clutch base plate - design change no. 2.

hence Rs. 4.55 + Rs. 4.95 or Rs 9.50 per piece. At 2000 pieces per month, the monthly savings potential is Rs 19,000.

2. The new design for the clutch plate Fig. 18.3 has circular holes instead of complicated serrated slots possibly reducing the procurement cost.
3. The 3 pins presently screwed on to the clutch can themselves be used to fix to the clutch pate. By this, the costly milling operation can be eliminated. The proposed design has 6 holes for better strength and power transmission.
4. The provision of nuts may involve slight modification on the crankcase, without effect on any functional design of the latter.

(c) **Supplementary design change no. 3: Reducing the bar size:** The bar procured is of 80 mm OD. It is provided with a 19 mm bore to match with the clutch shaft and is provided with two steps at 55 mm dia and 62 mm dia, the former to match with the clutch plate and the latter to match with the gear. Discussions with the R&D revealed that the additional step has no functional value and that the 62 mm step can be reduced to 58 mm. Thereby reducing the OD to 75 mm from 80 mm. Table 18.4 indicates the cost reduction achieved by reducing the bar sizes ignoring the reduced machining time.

(vii) **Follow-up phase:**

- Recommendation No. 1, viz to change to forging component, has been accepted and implemented. Net saving is Rs. 4.55 per piece or Rs. 9000 per month

TABLE 18.4 Cost reduction Achieved by reducing the bar sizes.

Sl. no	Radial overlap in mm	Bar dia required in mm	Wt. in kg per meter	Reduction in weight from 80 mm bar in kg/m	%age reduction in weight, thereby in raw material procurement price.
1	9	80	39.5	–	–
2	8	78	37.5	2.0	5.0
3	7	76	35.6	3.9	10.0
4	6.5	75	34.7	4.8	12.2
5	6	74	33.8	5.7	14.5
6	5	72	32.0	7.5	19.0
7	4	70	30.2	9.3	23.6

- Recommendation No. 2 viz Integration of the flange portion of the base plate with the clutch plate itself, by screwing, was accepted in principle but was deferred due to involvement of a change in the die of the aluminum casting of the crankcase. Correspondence was initiated with the supplier nevertheless without much positive result in this direction. Net saving is s 4.95 per piece or Rs. 25,000 per month
- Recommendation No. 3, viz reduction in the OD of the bar stock to an optimal level was withheld due to the design imposition from the principals. Expected savings Rs. 19,000 per month.

18.13 Conclusion

It can be seen from the case study that Value Engineering which has been practiced by industrial engineers even when the concept of Six Sigma did not exist, is still prevalent and has become a significant tool of Design for Six Sigma (DFSS).

The savings illustrated in the case study may appear to be too small to call for a citation. But in 1974 when this study was done, the petrol price was Rs. 1.50 per liter and rice was nearly a rupee per kg., compared to today's prices of Rs 85 and 60 respectively!

Criteria questions

1. What do you understand by Value Analysis? How it is different from Method study? (18.1)
2. Distinguish between SREDDIM and DARSIRI. (18.3, 18.10.7)

3. What are the different functions performed by a product? (18.8)
4. What are the different forms of value? How do you distinguish between functional value and cost value? (18.9)
5. Justify the statement 'Method Study: Production: Value Analysis: Design'.
6. List the phases of Value Analysis. (18.10)
7. Discuss the significance of function phase. (18.10.3)
8. What is DFSS and how does Value analysis achieve it? (18.1, 18.13)

Further reading

1. Otto K, Wood. *Product design.* Pearson Education; 2004.
2. Maynard HB, editor. *Industrial engineering handbook.* McGraw Hill; 1963.
3. ILO. *Introduction to work study.* ILO; 1979.
4. Ireson, Grant. *Handbook of industrial engineering & management.* Prentice Hall; 1971.
5. Hurst R. *Industrial management methods.* UK: Huchinson; 1970.
6. Yoga, krishnan A. *Project studies in industrial engineering*, vol. I. NPC; 1974.
7. Walker WF. *Engineering productivity.* BI Publications; 1965.
8. Eilon S. *Elements of production planning & control.* McMillan & Maruzan; 1962.
9. Kiran DR. Value engineering, a case study. *Ind Eng J* 1977.
10. Kiran DR. Total quality management, an integrated approach. BSP; 2017.
11. Kiran DR. *Production planning and control, a comprehensive approach.* BS Publications; 2018.
12. Kiran DR. Value engineering. *NIQR J* 2015.

Websites

1. http://www.businessdictionary.com/definition/value-analysis.html.
2. www.productivity.in/ValueAnalysis.
3. en.wikipedia.org/wiki/Value_engineering.
4. www.gsa.gov/portal/category/21589.
5. www.investopedia.com/terms/v/value-engineering.
6. www.dfma.com/resources/vave.htm?.

Chapter 19

Material layout planning

19.1 Introduction

Just observe the cobber who spends a lot of his time in shape fitting, that is planning out the shapes on the piece of leather available with him so as to get maximum number of pieces from his limited raw material. Same is the case with the tailor, with his cloth. The engineer likewise is concerned about the efficient preplanning of the material layout for the production of blanks from steel sheets, whether by punching on presses or by shape cutting on shearing machines, or even flame cutting so as to reduce the scrap production to a minimum.

19.2 Significance of material layout planning

A simple illustration given below emphasizes the significance of the material layout planning, before we go in for tool design, as cited in Chapter 3.

Let us presume that the blank to be produced is as per Fig. 19.1, with the dimensions as indicated. There can be three alternative layouts of the blanks as shown in Fig. 19.1B.

Assuming a minimum bridge of 2 mm between the blanks and the end near the edge, the strip sizes and areas in these three alternative layouts are as per Table 19.1, from which we can see that the alternative B has 24.1% less scrap than alternative A, while the alternative C has 20.5% less scrap. This fact influences a lot in the selection of the press capacity, strip size, tooling, etc.

19.3 Material layout planning applied to shearing operations

In addition to blanking by presses as illustrated above, the material layout planning is significant even when the blank is sheared from sheet metal by power guillotines. The case study below illustrates in detail a study undertaken for optimizing the material utilization of sheet steel, at a medium sized bucket manufacturing undertaking producing 1500 buckets a day.

Work Organization and Methods Engineering for Productivity.
DOI: https://doi.org/10.1016/B978-0-12-819956-5.00019-4

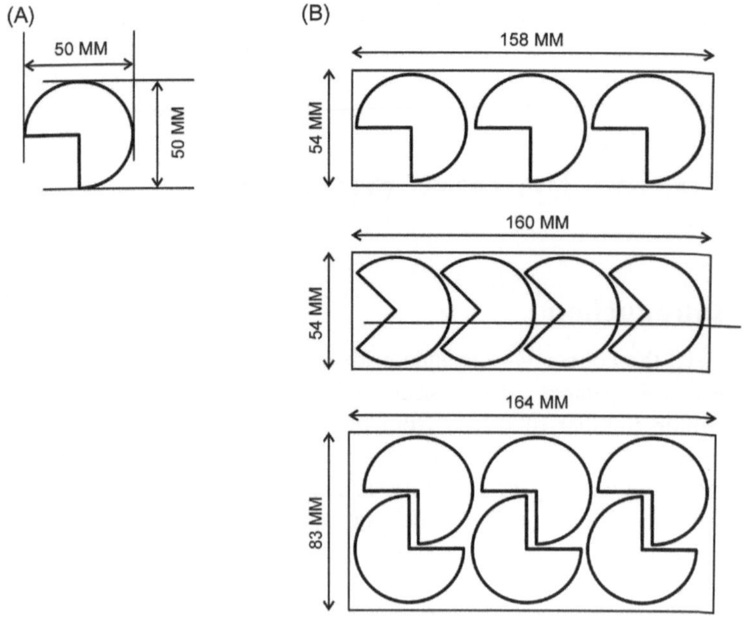

FIGURE 19.1 Alternative material layouts.

TABLE 19.1 Strip area for the production of the blank.

No.	Alternative	Feed (mm)	Strip width (mm)	Strip length per 1000 pcs (m)	Strip area per 1000 pcs (m^2)	%age reduction of scrap
1	A	52.0	54	158/ 3 = 52.7	2.848	–
2	B	37.8	54	160/ 4 = 40	2.160	24.1
3	C	54.0	83	164/ 6 = 27.3	2.265	20.5

19.4 Bill of materials (BOM)

The bill of materials (BOM) is a comprehensive list of the raw materials, sub-assemblies, intermediate assemblies, sub-components, parts and the quantities of each, required to build or manufacture an end product. BOM generally originates at the development or the design department and specifies the relationship between the end product (independent demand) and the components (dependent demand). It can be said to be the recipe and shopping list for creating a final product. It also gives an estimate of scrap that will occur during the production process, as seen in the case study below.

19.5 Case study for material layout planning

A certain medium scale industry produces 1500 galvanized steel buckets per day made from 22 g cold rolled steel sheets

19.5.1 Bill of materials

(a) Fan shaped body – 2 off - sheared from a 22 g cold rolled steel sheet in 30″ wide coils of 2500 kg. This forms 70% by weight of the raw material for the bucket.
(b) Bottom circle – 1 off- sheared from the same 22 g cold rolled steel sheet in 30″ wide coils of 2500 kg. This forms another 15% by weight of the raw material
(c) Base ring – 1 off- sheared from a 20 g cold rolled steel strip or sheet in coils.
(d) Ears – 2 off punched from a 16 g old rolled steel strip.
(e) Handle – 1 off - bent to shape from 10 mm MS rod.

The bucket has 5 components listed above and as illustrated in Fig. 19.2.

19.5.2 The bucket production process

The fan shaped bodies are first blanked as trapezoidal sheets as per process detailed above. The fan shape is then obtained by shearing in a nibbling machine. Two such bodies are joined by pressure rolling the seams and

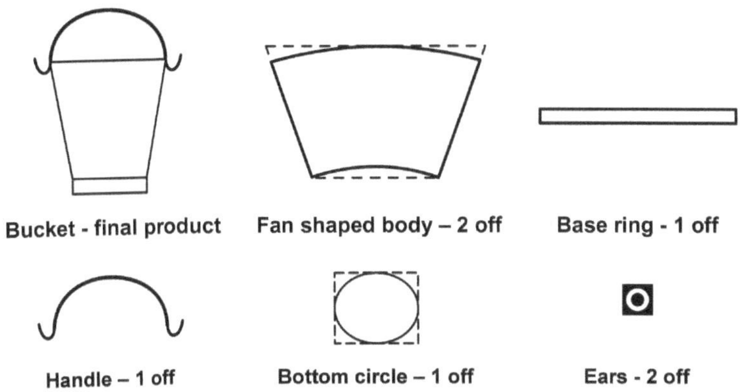

FIGURE 19.2 The basic components for the bucket.

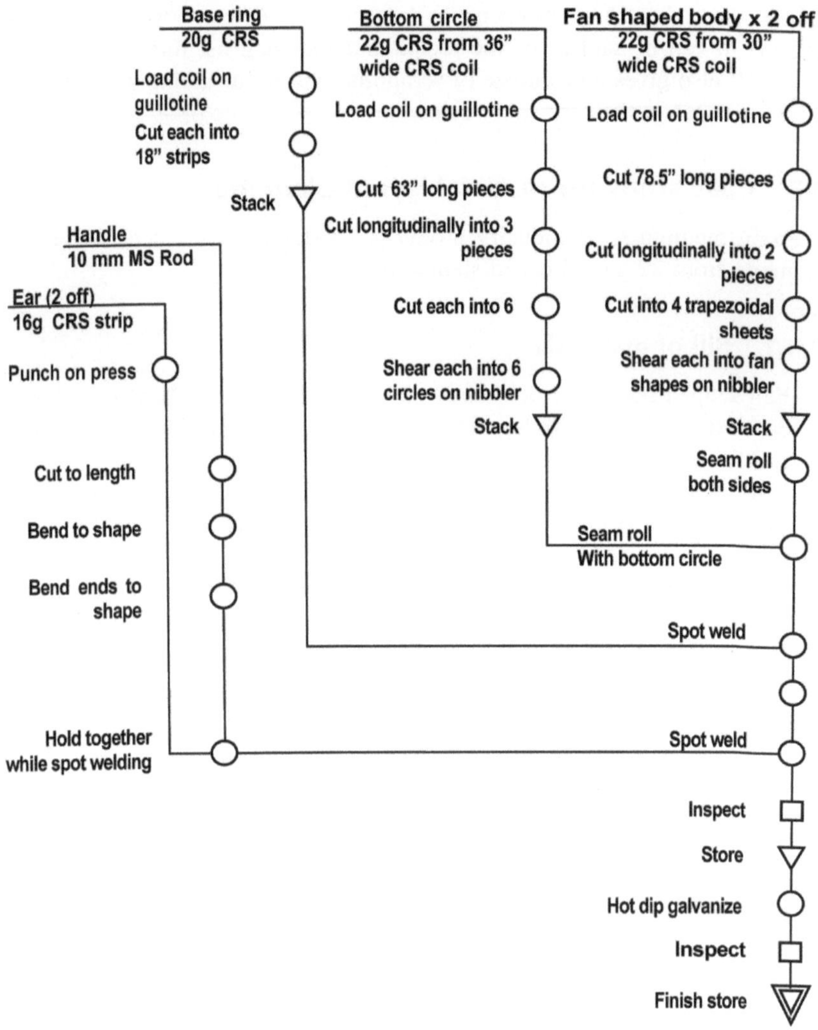

FIGURE 19.3 Operation process chart for the bucket production.

joined to the bottom circle (which is obtained by circle cutting of the 10.5″ × 11″ blank). The base ring is then spot welded and so are the ears with the handle duly inserted. The bucket is galvanized by hot dip method. The operations process chart as per Fig. 19.3, details the above process graphically.

19.5.3 Existing operation sequence for producing the blanks

Two machines, viz., the power operated guillotine and a foot operated manual shearing machine are used for the blank production.

A. For fan shaped body

(i) The 30″ wide CRS coil is at first sheared in the power guillotine into rectangular pieces of 30″ × 78.5″ till the coil is completed. The feed line of the sheet is perpendicular to the shearing edge.

(ii) Each of these sheets is again sheared longitudinally into 2 strips of 15″ × 78.5″ on the power guillotine.

(iii) Each of these strip is then cut into 4 trapezoidal blanks on the manual shearing machine leaving two triangular scrap pieces of 4.5″ × 15″

B. For bottom circle

(i) The 36″ coil is at first sheared on the power guillotine into rectangular pieces of 36″ × 63″. Here again the feed line is perpendicular to the shearing edge.

(ii) Each of the above strips is sheared longitudinally into 3 strips of 11″ × 63″, leaving a narrow strip of 3″ × 63″ that could be used elsewhere.

(iii) On manual sheering machine, each of these strips is cut into 6 pieces of 11″ × 10.5″

The total scrap produced during these shearing operations for the two components, excluding that during the nibbling, etc., subsequent operations would be 109.8 kg per 1000 buckets as detailed in Table 19.2.

19.5.4 Recommended material layout and the process

19.5.4.1 Change in the angle between the shearing edge of the power guillotine and the coil feed line

1. The axis of the de-coiler's shaft holding the CRS coil can be shifted by 16.5° so that coil feed is 73.5° to the shearing edge of the power guillotine. Alternatively the guillotine's position itself can be shifted by 16.5°. In fact the latter was preferred in the actual study since the working area occupied by the coil feed is more than the area occupied by the guillotine (Fig. 19.3)
2. The procurement of the CRS coil for both the fan shaped body and bottom circle can be limited to 36″, which is more easily available than 30″ wide coils. This would also facilitate the reduction of the coil sizes from 2 to 1 (Fig. 19.4).

19.5.4.2 Shearing of the fan shaped body in the recommended method (plan C of fig. 19.5)

1. The 36″ coil is loaded in the de-coiler. After shearing out a triangular portion of 11″ × 36″, which will occur only twice per coil of 2500 kg, subsequent shearing is at a spacing of 15.75″ to obtain parallelograms shaped bodies till the coil is completed.

TABLE 19.2 Comparison between the existing and proposed material planning layout.

Sl. No	Component	Plan	Shape	Scrap size	Frequency (per 1000 buck)	Scrap pet 1000 buckets			
						Reusable		Non-reusable	
						sft	kg	sft	kg
I Existing process									
1	Fan shape	A	4 Triangles	4.5 × 15″	4	–	–	234.4	109.8
2	Bottom Circle	B	1 Strip	3″ × 63″	18	72.9	34.2		
			Total scrap in the existing process			**72.9**	**34.2**	**234.4**	**109.8**
II Recommended process									
1	Fan shape	C	2 Triangles	11″ × 36″	1200			2.3	11.1
			1 strip	3/8″ × 15.75″	1			39.2*	18.4
2	Bottom Circle	D	2 Triangles	22″ × 8″	6000			0.2	0.1
			2 strips	0.5″ × 11″	6000			0.1	0.0
			6 Triangles	3.3″ × 11″	18			42.0	19.7
			1 strip	3′ × 88.5″	18	77.0	36.0		
			Total scrap in the recommended process			**77.0**	**36.0**	**83.8**	**39.3**

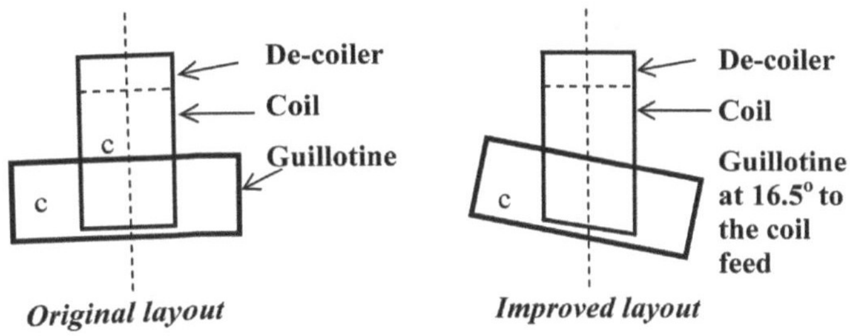

FIGURE 19.4 Change in the feed angle by shifting the Power guillotine.

2. Each of these parallelograms is cut on the manual shearing machine into 2 trapeziums of the required size. It may be noted that an excess of 0″ to 3/16″ is available depending upon the accuracy of the cutting process which would ensure that the cut size would under no circumstances be less than required and any excess would be included in the seamed portion of the seam rolling.

19.5.4.3 Shearing of the bottom circle in the recommended method (plan D of fig. 19.5)

1. Just as above, the 36″ coil is loaded in the decoiler of the power guillotine. After shearing out a triangular portion of 11″ × 36″, which will occur only twice per coil of 2500 kg, subsequent shearing is at a spacing of 66.5″ to obtain parallelograms shaped bodies till the coil is completed.
2. Each parallelogram is sheared into three strips of 11″ width, leaving a reusable scrap strip of 3″ width.
3. Each 11″ strip is cut on the manual shearing machine into 6 rectangular pieces 10.5″ × 11″ leaving two triangular scrap pieces.
4. It may be noted that from the first and last trapezium of the coil, an additional bottom circle each can be obtained. That means 19 bottom circles can be obtained from the first and last trapeziums while the subsequent parallelograms would yield 18 bottom circles each.

The total scrap for these operations for both components in the recommended process is only 39.3 kg per 1000 buckets against the original scrap of 109.8 resulting a reduction of 64.2% by weight of the scrap.

19.5.5 Summary of results achieved

1. By shifting the position of the guillotine the benefits are significant.
2. The scrap is reduced from 109.8. Kg per 1000 buckets to 39.3 kg per 1000 buckets or by 64.2%. In monetary terms the savings are Rs 250,000 per year at a production rate of 1500 buckets per day.
3. Material procurement has eased by elimination of 30″ wide coils.

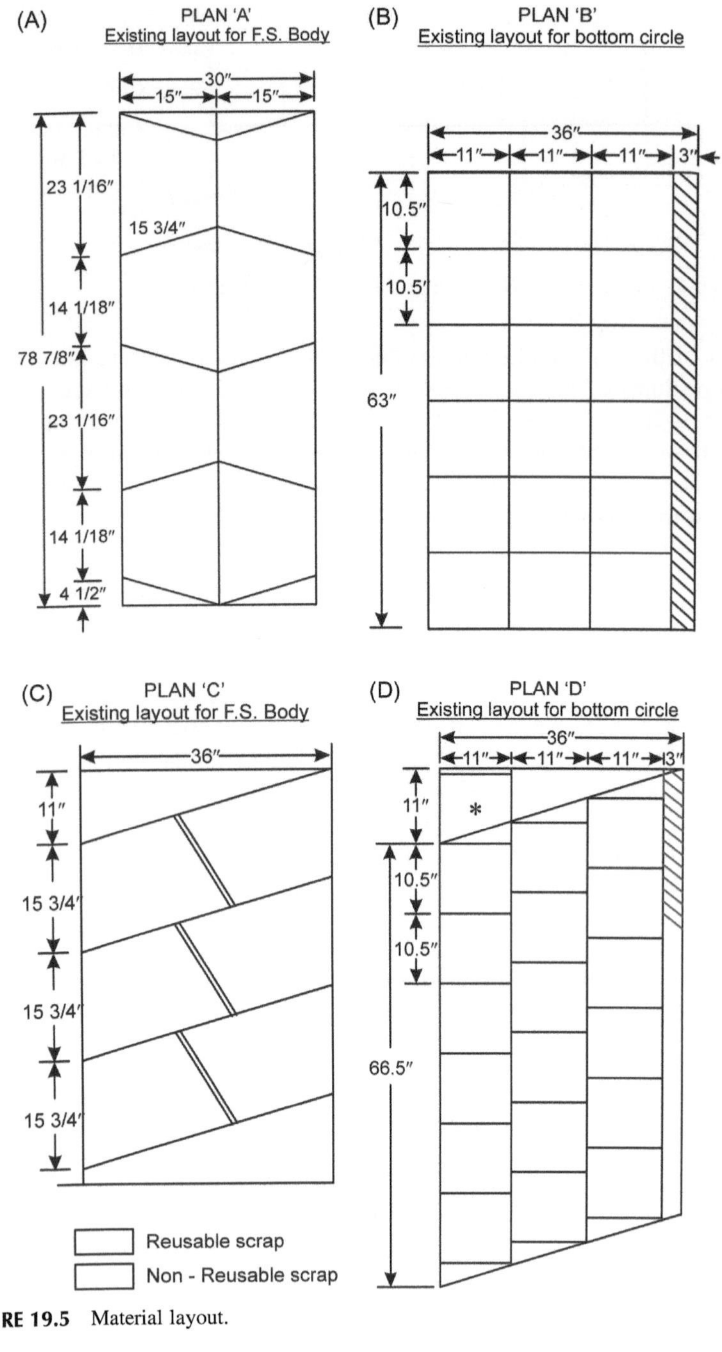

FIGURE 19.5 Material layout.

4. The total operational time for these 2 groups of operations for the two components of fan shaped body and bottom circle has been reduced from 1.858 man minutes to 1.468 man minutes per bucket corresponding to an increased output of 11.5%.
5. The scrap produced in the subsequent operations like nibbling and ear punching are not considered in this study as they remain unaffected by the recommended process.
6. The shifting of the guillotine has little or no effect on the production of base rings or ears. The new position of the guillotine does not affect the workplace layout or the aisle space.

19.6 Conclusion

As emphasized in the introduction, new product development is a significant requirement for organizations to survive in this buyers' market. The various features described in this chapter would enable the production planning and control engineer to understand the need for the close coordination needed with the R & D department. The case study on material layout planning would help the production planning engineer apply creativity in improving processes.

Criteria questions

1. Discuss the Significance of material layout planning. (19.2)
2. What is Bill of Materials? How does it help in production planning? (19.3)
3. In the case study, the guillotine is turned by 16.5° and refixed. Why not the coil unwinding mechanism? (19.4)

Further reading

1. Kiran DR. Professional ethics and human values. McGraw Hill India; 2013.
2. Kiran DR. *Total quality management: an integrated approach.* BS Publications; 2015.
3. Kiran DR. *Production planning and control, - a comprehensive approach.* BS Publications; 2015.
4. Kiran DR. *How to be more creative.* House Journal of Ralli Group; January 1975, p. 6 & 7.
5. Kiran DR. Re-layout of guillotine for the bucket body shearing operation - a report submitted to M/s metal products. Dar Es Salaam, Tanzania: ALAF group; 1976.
6. Kiran DR. *Value engineering: a case study - Industrial Engineering Journal,* December 1977, p. 30–33.
7. Kiran DR. *Material layout planning . Ind Eng J* 1980; 11–16.
8. https://en.wikipedia.org/wiki/Product_design.
9. www.businessdictionary.
10. http://www.cyric.eu/.
11. Lidwell W, Holden K, Butler J. *Universal principles of design.* Rockport publishers; 2003.
12. Kelley T. *The art of innovation: lessons in creativity.* Jaeper Collins Business; 2001.
13. Ulrich KT, Eppinger SD. *Product design and development.* 5th ed. McGraw Hill; 2003.

Chapter 20

Work study on clerical operations

20.1 Introduction

The importance of work study in general is traditionally associated with a manufacturing environment with a purpose of improving the efficiency of manufacture of goods from resources. Nevertheless, its importance of resource utilization within the office becomes ever more significant due to the growth of the administrative and clerical costs in most organizations. These must be controlled, and work study is the foremost tool for achieving this.

The basic problem in measuring employee performances is that no established method of counting or recording the input of clerical staff exists. In short, I.L.O stresses that

> *"A reason that work study is often not practised within offices is the different status of manual and clerical workers. What is perceived as a factory technique is not acceptable to those who think of themselves as having higher status."*

20.2 Organization & methods

Though our common precepts on Organization & Methods is that of the management of the administrative operations, most definitions on O & M, cite it as the systematic examination of activities in order to improve the effective use of human and other material resources. We will use its nomenclature as a synonym for the method study applied to clerical and administrative operations.

It is normally a consultative service to management, since the line management does not have the time nor awareness of priority to conduct this analysis in office work. It is sometimes called as Business Improvement, Business Analysis, Project Management, Process Improvement or Internal/Business Consultancy.

A.F Leaman, in his book Social Security Administration, asserts

Work Organization and Methods Engineering for Productivity.
DOI: https://doi.org/10.1016/B978-0-12-819956-5.00020-0

Administrative reforms are an effort to mitigate the gap between reality and what is desirable of the administration. Work study is the solution for it. As a management aid tool, work study helps in work simplification and efficiency of management processes. Administration or to be very specific, the Bureaucracy from a traditional point of view has always been change resistant, opaque, and rigid. Therefore, administrative improvement and reforms are a priority.

The office is the administrative center of a business where all sorts of clerical work is done to co-ordinate and control the affairs of the whole organization. William H.; Robinson, Edwin M. Leffingwell, in their book, Textbook of Office Management, state that

Office work is concerned with records and statistics, with communication, with computing, with planning and scheduling. Every office task comes within the scope of one or the other of these activities.

Office management could be regarded as an act of putting into systematic relationship, those elements and activities essential to the satisfaction of an office purpose. It has no reference to size or nature of activities that are taking place in any Organization, whether it is a room where lecturer attends to his students, or where the president of United States carries out his constitutional duties.

The following, on the part of the office staff of an organization, are some of the factors that may lead to low employee performance and even loss of documentation

- Lack of the procedural knowledge of the processes of record keeping and documentation in the office,
- Lack of the priority given by the employees on desired manner in which office should be kept, and
- Lack of the awareness of organization and methods.

20.3 Definition of organization and methods (O & M)

Organization and methods is the systematic examination of activities in order to improve the effective use of human and other material resource.

British standard 3188.1001.

Systematic examination of an organization's structure, procedures and methods, and management and control, from the lowest (clerical or shop-floor) level to the highest (CEO, president or managing director). Its objective is to assess their comparative efficiency in achieving defined organizational aims. O & M concerns itself mainly with administrative procedures (not manufacturing operations) and employs techniques such as operations research, work-study, and systems analysis.

http://www.businessdictionary.com/definition/Organization-and-Methods-O-M.html.

Organization and methods is a specialist function which has a primary objective of improving efficiency and control in organizations.

http://www.managers-net.com.

Organization and Methods is the assessment of organizational hierarchy, policies and procedures, and management systems. The aim is to identify organizational inefficiencies, relative to rival organizations, usually within administrative functions.

Black's Law Dictionary.

Organization and Methods is the process of examining how the work is done in a business, industry, etc and finding ways of doing it more effectively.

http://lexicon.ft.com.

Organization and Methods is the process of examining the structure and working methods of a company or organization in order to make it more effective.

Cambridge Dictionary.

20.4 Application of work study in office

SREDDIM, the basic procedure of method study is valid for an office environment too, as illustrated in the following steps.

20.4.1 Select

In the office, as cited by ILO, we shall be looking for areas or activities that:

- account for a significant proportion of office labor costs (Pareto analysis can effectively be used here);
- are producing large numbers of errors or serious errors;
- are creating high levels of dissatisfaction;
- introduction of new technology such as computerization that necessitate changes in the clerical responses to some external change;
- As and when people change, short cuts in form filling, etc. will be made, thinking that the figures are self-explanatory;

In a nutshell, work study does not achieve as much reduction in clerical costs as it does for reducing manufacturing costs.

20.4.2 Record

Documents and forms are common features of all offices, being the carriers of information and sometimes being the only physical evidence of a procedure. Whereas in shop floor the flow of material or the end products etc., which are physical, are recorded, in office the recording made is generally about some information or other, written on either white paper or on roughly self-designed forms, as per the interpretation of the recorder, which is generally different from that of the reader.

This practice, hence brings to the forefront the most essential feature of the office documentations, the forms design and control of the forms. Even after form design, occasionally many new forms arrive on the scene, but very few forms are ever deliberately, discontinued. Hence there is an equal need for the control of forms.

One of the best and most effective form for the analysis of clerical workload is the work distribution chart which is a multicolumn document flow chart. The Document Flow Chart aims to observe and record how the documents flow from person to person in any business situation. Documents such as orders and invoices usually have many copies. They affect other documents and then some other documents, until eventually they are filed or sent to a customer. People can do a number of things to documents. They can: check, collate, separate, add to, and use to originate other documents. Sometimes they can have several choices. All these actions are recorded in this multicolumn document flow chart, which will facilitate the application, of critical evaluation or brainstorming in reduction of the number of copies or the flow pattern itself.

20.4.3 Evaluate

The following steps forms the procedure for the evaluation of the recorded data made during the earlier stages.

Step 1 Critical questioning- 'What and Where and When and Why and How and Who',

As in method study, every form should be examined critically likeForms designs

1. Is the form necessary?
2. What information does it convey?
3. Who uses it?
4. When do they use it?
5. Where is it used?
6. How is it used? (Filled in manually or typed or computer printed, etc...)

Step 2 Designing or re-designing the form

1. Check compatibility with its intended use.
2. Identify the nature of information to be filled in and degree to which they can be grouped.
3. If it is simple and easy to fill in.
4. Whether it is Easy to understand and interpret the information.
5. If it can be interpreted by all in the same manner, and
6. How many copies to be made and to whom.

Step 3 Detailed design of usage and storage

1. The filing process,
2. Data retrieval process,
3. Routing the form through the relevant departments.

Step 4 Control of the forms

This is the most important stage, as explained before.

1. Regular audits to be conducted to check if each form still serves its purpose.
2. Maintain a central register of all forms in use, together with a review schedule for each form.

The above steps of evaluation are expected to reduce unnecessary work in

1. Doing another department's work.
2. Duplicating the work done by others as well.
3. Spending too much time on unimportant jobs.
4. Having skilled people doing unskilled work.
5. People doing unrelated work.
6. Unbalanced loads between employees.

20.5 Obstacles to administrative reforms

1. Excessive tolerance to maladministration.
2. Political and bureaucratic inertia.
3. Lack of funds and infrastructure.
4. Lack of research, knowledge and opposing ideas/ideologies, etc.
5. When these reports on the same are considered to idealistic and not implementable practically.

20.6 Avoid cluttering of office desk

This is very important in organizing the office or methods, and an ideal case for conducting methods study on clerical staff. Fig. 20.1, a cartoon from the New York Times Magazine of 1964, illustrates how a badly organized office desk would cause loss of working efficiency.

20.7 Conclusion

Though according to I.L.O, work study of clerical operations is difficult, it is not impossible. Several O & M studies have been undertaken to improve the working methods of clerks. As an illustration, this author had conducted a study on the movement of bale slips during the different processes of cloth

FIGURE 20.1 A bad style of desk working.

weaving and baling at Khatau Mills, Mumbai, and achieved considerable reduction of the work load of the clerks.

Criteria questions

1. What is implied by O & M? How it is relevant to work study on clerical operations? (20.2)
2. Explain the factors that lead to low employee performance in office administration. (20.3)
3. Discuss the application of SREDDIM in office operations. (20.4)
4. What is meant by 'forms design' and how do you control them? (20.4)

Further reading

1. ILO. *Introduction to work study*, 3rd ed. ILO; 1979.
2. Leaman AF. *Social security administration*. USA: Social Security Administration; 2017.
3. Robinson EM, Leffingwell WH. *Textbook of office management*. 2nd ed. McGraw Hill; 1993.
4. Kiran DR. *Report on the distribution of Bale Slips in Khatau Mills*; 1972.
5. http://www.managers-net.com.

Chapter 21

Resistance to change

21.1 Improvement vs resistance

We all know that most of the management techniques to raise industrial productivity aim at method improvement or systems improvement. We also said that work study is a penetrating tool of management, opening up the inefficiencies in the system. The very word 'improvement' is linked with an action of change.

In the day to day shop meetings at the factory manager's level, if the workers or their representatives are invited to participate along with the departmental heads, the employees would feel responsible for the job in view of their participation. This is called participative management. On the other hand, if the management decides to implement certain changes or improvements without consulting the workers or their representatives, there would be certain amount of resistance from the workers, even though those improvements would ultimately benefit the worker. This is called Resistance to change and is experienced by a majority of the industrial engineers working in an industry.

In most of the cases where a resistance is encountered to a change, the manner in which the implementation is planned and conducted, with due regards to the ego of the concerned worker, and the manner in which he is taken into confidence in assisting the change and in the process of implementation, is the very key to the success for any continuous improvement project.

While in a majority of cases, the resistance is from operatives, many a times, the resistance comes from personnel, higher up in the ladder like the departmental heads also.

A sincere attempt to understand the possible motive for such a resistance, as well as the past history responsible for the development of such a resistance, together with the complete analysis of the situation would possibly enable the management to plan a rational course of action for the successful implementation of the change. The case study given at the end of this chapter (paragraph 21.18) illustrates this.

21.2 Types of changes that generally meet resistance

(a) Changes that threaten to lower the prestige of worker of a group or of an individual like an office clerk being asked to work within the workshop.

Work Organization and Methods Engineering for Productivity.
DOI: https://doi.org/10.1016/B978-0-12-819956-5.00021-2

(b) Changes that highlight the inefficiency in the present procedures, resisted mostly by higher ups.
(c) Changes that reduce the authority and scope of decision. If the personnel department is asked to scrutinize or to take control over the constant hiring or firing of the workers, the foreman might feel the loss of his importance.
(d) Changes that interrupt the routine work, like the re-layout projects.
(e) Change of processes, when the workers resist due to the fear of loss of employment or reduction in earnings.
(f) Changes involving development of skills in new directions.
(g) Changes involving transfer to new environments, like moving from a cozy room to a large hall or moving to a new set of workers.

21.3 Effect of worker representation on productivity

Fig. 21.1 illustrates the effect of Worker Representation on Productivity

21.4 Reasons for resistance

When viewed from the worker's point or view, they do have their own valid reasons to resist a change. A few commonly cited reasons put forth by workers or their representative may be enumerated below:

(a) New method may result in retrenchment, and he may be the one among those affected.

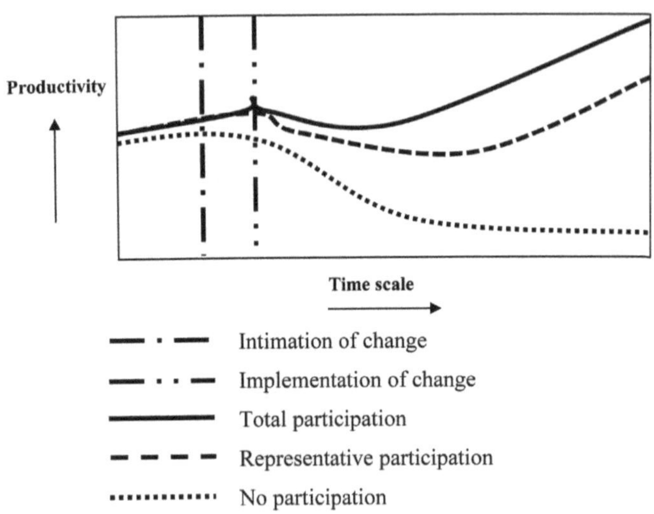

FIGURE 21.1 Effect of worker representation on productivity.

(b) He may not be proficient in the new job, resulting in the reduction of the wages earned.
(c) The experience gained in the previous job over years, would be reduced to a waste.
(d) He has to create a new circle of friends under a new supervisor, which may or may not suit him.
(e) Psychological attachment to the place or machine or the group with which he worked for years.
(f) Fear of derecognition caused by insufficient information received by him about the purpose for the change, resulting in distorted rumors.
(g) Though he is not against the change personally, he may be bound by the decision of the union or the group to which he belongs to.
(h) Last but not the least, his ego might have been hurt, for not being consulted before the change. This fact could be understood clearly by the graph, shown in Fig. 21.1.

Of the above, the last cited reason has created a revolution in the management thinking, appreciating a need for involving the workforce not only in the improvement projects, but also in the day to day management, suggestions, etc., which concept is generally termed now as Worker Empowerment.

21.5 Some criticisms generally encountered in the process of change

Sometimes the management is bent upon enforcing change. A sense of despair creeps in amongst workers all round. Such situations give rise to a number of criticisms from those affected persons either in hush-hush tone or they air their criticism to the management as well. Some of them which are commonly treated are:

- "We tried that before"
- "Our place is different"
- "We don't have the time"
- "Not practicable for the operating staff"
- "Let us get back to realities"
- "Why change it? It is still working all right"
- "Good thought, but not practical"
- "Don't you realize that the present one is developed after so much research considering all the factors you are pointing out now?"
- "We already thought of this and we know it cannot be done"
- "We already tried this, but it was a flop"
- "Don't you think we too get good ideas?"
- "Who are you to tell me? I know my job"

21.6 Employee involvement strategies

The following are some of the strategies to ensure the involvement of the employees.

Suggestion Schemes: Programs that encourage individual employees to put forth their ideas of improvement in the process or any operation or work environment.

Survey Feedback: Use of employee surveys as a part of the larger problem-solving process in which the survey data is used to encourage, structure and measure the effectiveness of employee participation.

Quality Circles: Groups of shop floor level, meeting voluntarily and regularly in a structured environment under the guidance of the supervisor to identify and suggest work related improvements. In view of the significance of this concept in the recent decades, this has become a major management tool, in improving quality productivity, etc., that every organization has a quality circle. It is estimated that several millions of operatives are members of these quality circles, worldwide. Even in India, the Quality Circle Forum of India has become a very active forum, well recognized in industrial circles.

Job Redesign: This involves the three elements, job rotation, job enlargement and job enrichment.

Self-Management Teams: Similar to the quality circles, but with a definite assignment like safety incorporation in specific process. This may also include certain support functions like hiring of personnel for that assignment.

Employee Participation Group: This involves all levels of employees, convened by the management for a particular task.

Motivation: Motivation is the basic psychological factors in making an employee feel himself as a part and participle of the company and getting him to work with his full heart and soul on the job. This has been a subject of interest right since man started living as a social animal. Several theories have been developed to understand this psychological factor. The following are some of the theories propounded on motivation.

21.7 Abilities of man vs machine

As we have seen in Chapter 8, we can say that the human brain has many powers.

- Absorptive power - the ability to observe and apply attention
- Retentive power - the ability to memorize and recall
- Reasoning power - the ability to analyze and judge
- Creative power - the ability to visualize, for ease and generate ideas.

By first two we learn and by the latter two, we think. A computer may be able to perform the first three better than the human, but the creative power is the specialty of human brain.

Highlighting this concept in his book *Human Factors in Engineering and design*, McCormick makes the following generalization about the relative capabilities of humans and machines, and states that humans are generally better in their abilities to

1. Sense very low levels of certain kinds of stimuli, visual, auditory, tactual, taste, etc.
2. Detect stimuli against high background disturbance, like the bleepson TV screens with poor reception.
3. Recognize patterns of complex stimuli which vary from situation to situation, like aerial photographs and speech sounds. Thereby sensing unusual Sense unusual and unexpected events in the environment.
4. Store (remember) and retrieve (recall) to some extent, the information over long periods of time.
5. Draw upon varied experience in making decisions, adapt the decisions to suit situational requirements.
6. Select alternative modes of operations if certain modes fail.
7. Reason inductively generalizing from observations.
8. Apply principles to solutions of varied and unexpected problems,
9. Use creativity to develop entirely new solutions.

In these days of robotics and Internet of Things, machines are designed to be smart with embedded vision sensors, and with deep learning principles, by which they are programmed to recognize speech, text, photos, etc., to excel humans in several controlling aspects (reference may be made to Chapters 34 and 35 of the text book on *Production Planning and Control: A Comprehensive Approach*, by this author). Nevertheless, it may be concluded that creativity is the sole specialty of human brain compared to machinery as cited in paragraph 8.3, and it is the human mind and the psychological factors that play a crucial role of improvements and changes.

21.8 Maslow's theory of hierarchy of basic needs

As seen in paragraph 21.6, motivation of the employee plays a vital role in his conceiving a change with an open mind. Thus, before discussing how to successfully implement a change, we should understand the psychology of the operative and why he fears changes.

Abraham Maslow has propounded that man is motivated mostly by fulfillment of his basic needs, which keep varying as time passes and as he achieves higher levels during his career. Once the lower level of needs is satisfied, he is no more motivated by that need. He wants to go a step higher. As illustrated in Fig. 21.2, as man starts his life, his basic need would be

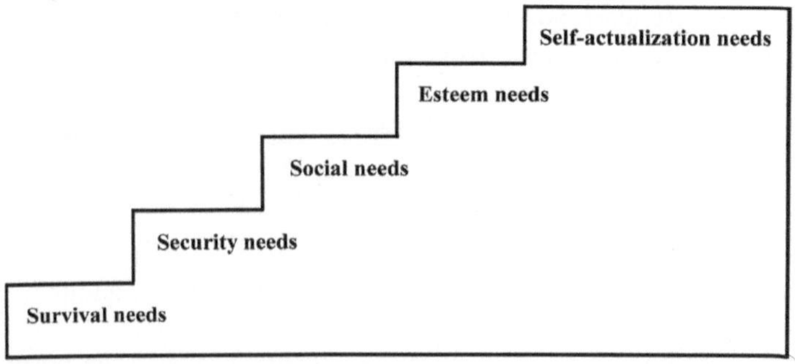

FIGURE 21.2 Maslow's hierarchy of needs.

food, shelter and clothing for survival. That means he just wants to earn sufficient money to feed on himself and his family. Once this is satisfied, his need would then be to have a secured job and life. This level includes a satisfying job and a safe working environment. At next level, he needs to satisfy his ego or social need that is the need to be part of a society around him. Within the organization he would like to be a part of the group of workers with whom he can have day to day free exchange of thoughts. Once this is satisfied, his ego or esteem would surface and he wants to be recognized in the society as someone important. At the self-actualization level, he wants to and be appreciated as a man of achievement. He wants to be given opportunities to go as far as his abilities can take him.

21.9 Theory X, theory Y and theory Z

During then 19th century, Sigmund Freud (1856–1939), emphasized the importance of the unconscious mind which governs behavior to a greater degree than people consciously think. Later, Douglas McGregor propounded that humans are diverse in their psychological outlook on life and categorized them into 2 groups under Theory X and Theory Y.

Theory X - This theory states that most people in the workplace do not enjoy work and will take every opportunity to avoid doing their job because they are lazy and need to be closely supervised, threatened and disciplined by management.

Theory Y - This is opposite to Theory X. Theory Y can be stated as a method of managing people in the workplace based on the idea that most workers enjoy their job, are self-motivated and want responsibility, and the managers' role is to help the workers realize their full potential by giving them more responsibility, including them in decision-making, etc.

Theory Z - On the other hand, William Ouchi developed a Japanese management style terming it as theory Z, which states, that workers like to build relationships with other workers and management, to feel secure in their jobs, develop skills through training, and have their family life and traditions valued.

21.10 How to successfully implement a change?

Always bear the following principles in mind, before planning for a change. This is the key to your success, given the task of planning and implementing an improvement project.

1. Understand the change itself, its benefits to the worker, its benefits to the company etc. enlist them. You personally be convinced of these benefits.
2. Predict the reasons for possible resistance from the worker's point of view. Be prepared with your solutions for these factors and other criticisms.
3. Explain the workers fully and convince them that the change is for their good only. Allay their fears.
4. Do not forget the channels of authority. Explain the concerned foreman first.
5. Listen to their suggestions and make a note of them. Incorporate them where possible.
6. Discuss with them freely and sincerely.
7. Create interest in them, by visual aids, manner of speaking, etc.
8. Give due consideration for their present skills.
9. Plan their training on the new jobs fully and systematically.
10. Keep asking them now and then, how they get along with the new job.
11. Commend them or correct them where necessary.
12. Last but not the least, review the proposal and make a list of the reasons for possible resistance from the worker's point of view and other possible criticisms from different levels and personnel. Be prepared with your solutions. This is as illustrated by the Japanese trem 'Dakara Nani?' in Chapter 12.

21.11 Empowerment

Empowerment is defined as the basis of giving employees skills, resources, authority, opportunity, motivation, as well as holding them responsible and accountable for outcomes of their actions, which will contribute to their competence and satisfaction. It is an environment in which people become stronger and more confident, especially in controlling their field of activity and take the responsibility and ownership to improve the process and take appropriate steps to satisfy the customer requirements and achieve organizational values and goals. It is a philosophy that bestows on the employees the authority and responsibility for taking decisions which affect their jobs. It also signifies the process of employee participation in company working practice and organization performance.

As indicated above, the conditions needed to be satisfied for empowerment are:

- The capacity for the employee to take the right decision,
- The commitment to take the responsibility, and
- The confidence and trust bestowed upon him by the management.
- Appropriate recognition and rewarding by the management.

Since empowerment largely involves in people taking decisions, Chapter 5 elaborates the process of decision making.

21.12 Benefits of employee involvement

Besterfield cites the following benefits of employee involvement and participative management:

1. Employees make better decisions using their expert knowledge and skill in the process.
2. Employees are more likely to implement and support decisions in which, they had a part in making.
3. Employees are better able to spot and pinpoint areas for improvements.
4. Employees are better able to make immediate corrective action.
5. Employee involvement reduces the labor/management friction by encouraging more effective communication and cooperation.
6. Employee involvement increases the morale by creating a feeling of belonging to the organization.
7. Employees are better able to accept change because they control the work environment.
8. Employees have an increased commitment to unit goals because they are involved.

21.13 Total employee involvement

Peter Grazier puts the following points in favor of Total Employee Involvement.

1. Everyone has something to contribute and will, if the environment is right.
2. The human element performance is more important than the technical element.
3. Most decisions can be significantly improved through collaboration.
4. People need Leaders. Good leaders build trust and a higher sense of mission and source of worth.

5. Employee involvement is not a program. It is a corporate philosophy.
6. Continuous improvement is beautifully simple. As improvements begin to flow, confidence will build and the process will feed on itself.

21.14 Recognition and rewards

This is the fourth component of employee involvement as cited in paragraph 21.11. Recognition is a process by which Management acknowledges the good performance of an employee. This is based on the esteem need as indicated by Maslow in his hierarchy of human needs. We had already seen that recognition and satisfying the esteem need of the worker is a major contributor for his motivation. This sustains the employee's interest and commitment in moving towards the common goal. Employee recognition is not just a nice thing to do for people. Employee recognition reinforces and rewards the most important outcomes people create for your business. In several occasions the recognition and verbal appreciation would boost the employee morale more than giving the cash awards. The rewards can either be intrinsic like the non-monetary rewards and extrinsic like the monetary rewards, as cited in the next paragraph.

21.15 Forms of recognition and rewards

The following can be some of the methods of recognition and rewards.

(A) Intrinsic rewards

- Verbal appreciation. Supervisors can give on the spot appreciation in front other employees.
- Certificates or plaques.
- Letters of appreciation from the Chief Executive Officer, Chairman, etc.
- Displaying the names on the notice boards.
- Other non-monetary rewards like inviting for a get-together or family dinner.
- Group incentives like departmental picnics, departmental annual holidays. Khatau Mills have the convention of sending their senior officers annually along with their families to holiday resorts like Goa, Mahabaleshwar, etc.

(B) Extrinsic rewards

- Profit sharing.
- Gain sharing.
- Employment security.
- Benefits of recognition and rewarding systems.
- Monetary rewards like cash rewards in the suggestion schemes.

- Job related incentives like double increments, promotions, cash bonus, gain sharing, etc.
- Productivity based incentives.
- Quality based performance bonuses.

21.16 Criteria for effective recognition of employees

The website https://humanresources.about.com/od/rewardrecognition/a/recognition_tip.htm cites the following tips for effective recognition of employees for establishing criteria for what performance or contribution constitutes rewardable behavior or actions.

- All employees must be eligible for the recognition.
- The recognition must supply the employer and employee with specific information about what behaviors or actions are being rewarded and recognized.
- Anyone who then performs at the level or standard stated in the criteria receives the reward.
- The recognition should occur as close to the performance of the actions as possible, so the recognition reinforces behavior the employer wants to encourage.

21.17 Advantages of effective rewarding systems

1. It is an effective employee motivator by letting them know that they are valuable members of the company.
2. Better and committed involvement of the employees can be ensured.
3. It also motivates other employees.
4. It creates a healthy competition among individuals and teams.
5. It increases morale in the company.
6. It provides a specific goal to the employees.
7. It provides the organization an opportunity to thank high achieving people.
8. It gives publicity to the company that they value quality and productivity.

21.18 Case study

The works manager of M/s XYZ & Co, a medium scale industry called the supervisors of all the departments for a meeting. An MBA from Harvard University, he opened the meeting with an appreciation on the good performance the company made in the past 2 years and commended the role of the workmen, especially the supervisors in achieving this high performance level.

He recalled how the company started its operations 10 years back in the old barrack type building. The office is located in the front portion while each of the 6 to 7 rooms housed stores and machinery, as and when they were procured. A small extension to this formed a semi open asbestos roofed space, housing maintenance, fettling, bench drills etc. A large backyard remained unused.

Due to the good performance of the company, expansion of activities became inevitable and it was contemplated to extend the building to cover a major portion of the backyard. The existing layout had obvious inefficiencies, causing lots of obstructions for free movement of men and materials. A comprehensive study was made with specific reference to the movement, and the improved layout, the works manager explained, took care of all these factors with more facilities for the workmen. The basic changes, according to him are insignificant excepting movement of certain machinery and removal of certain partition walls for easier materials handling.

All the supervisors stared blank towards the works Manager and showed neither the signs of happiness nor that they were overawed. But one thing was certain. They all knew their workers would resist the change. In the existing system of working, senior workers of machining department were closeted in a cozy little room, whereas in the revised layout they would be working in a large hall with others. In the hall there would be unskilled workers too. In the assembly department, 8–10 workers used to cluster around a large bench and work in quiet conditions. Their being shifted to the large hall would mean lot of noise and working on different benches separately.

All this meant too much for senior workers - skilled workers equated with unskilled ones, noise, no familiar faces around to talk to, etc. they decided to approach their union, who in turn, could dissuade the management not to undertake the re-layout work.

How would you have handled the situation?

The above was just one case in point to illustrate how resistance to change occurs The layout improvement project cited in our case study was planned very meticulously providing several new facilities to the workers, elimination of crowded work areas etc. this was projected as worker friendly and was expected to be implemented without resistance. But it was not so. If you are the works manager, how would you have handled the situation?

21.19 Conclusion

In a majority of cases where a resistance is encountered to a change, the reason is more psychological rather than the worker being against the change itself. A sincere attempt to understand the possible motive for such a resistance as well as the past history responsible for the development of such a resistance, together with complete analysis of the situation would possibly enable the management to plan a rational course of action for successful implementation of the change.

Hence the manner in which the implementation is planned and conducted, with due regard to the ego of the concerned worker, and the manner in which he is taken into full confidence in assisting in the change and in the process of implementation, is the very key to the success for any improvement project.

Criteria questions

1. Which types of Changes that you think generally meet resistance? (21.2)
2. What are the three kinds of employee participation? Illustrate their effect on productivity. (21.3)
3. Analyze the reasons for Resistance from the employees for change. (21.4)
4. Discuss the several strategies for involving the employee in change management. (21.6)
5. What is Maslow's theory of hierarchy? Illustrate. (21.8)
6. Explain Theory X, Theory Y and Theory Z. (21.9)
7. Why are recognition and rewarding necessary for successful change management? (21.14)
8. Distinguish between Intrinsic and Extrinsic Rewards. (21.15)

Further reading

1. International labor organization. *Introduction to work study*. 3rd ed. ILO; 1979.
2. McCormick. *Human factors in engineering and design*. McGraw Hill; 1993.
3. Kiran DR. *Production planning and control, a comprehensive approah*. BS Publications; 2018.
4. Kiran DR, editor. *Proceedings of the management seminar for transport sector*. Dar Es Salaam: National Institute of Transport; 1984.
5. Kiran DR. Participative management, a case study. *J Indian Inst Prod Eng* 1999. Jan.
6. Kiran DR. *Total quality management: key concepts and case studies*. USA: Elsevier; 2018.
7. Kiran DR. *Maintence engineering and management - precepts and practice*. USA: Francis and Taylor; 2016.
8. Kiran DR. *Professional ethics and human values*. McGraw Hill Higher Education; 2012.
9. Maslow AH. A theory of human motivation. *Psychol Rev* 1943?.
10. https://en.wikipedia.org/wiki/Maslow_hierarchy_of_needs?.

Chapter 22

Industrial engineer's role as a consultant

22.1 Who is a consultant?

A consultant is a qualified person providing services independently. in identifying and investigating into management problems that arise in the procedures, methods, policy, etc., of an organization, thereon recommending appropriate actions and helping in the implementation of these recommendations.

Almost every industrial engineer in his day to day work may have to conduct project studies, whether he is a methods engineer, production engineer or maintenance engineer, and report to his boss either verbally or by routine reports or by project reports. In this manner, every engineer is an internal consultant of his organization, and hence needs to know more about effective communication.

22.2 Key features of consultancy

- Consulting is an independent service.
- It is in an advisory role.
- It provides specialized knowledge and skills.
- It provides practical solutions for problems.

22.3 Why are consultants used?

- Providing special knowledge and skill
- Supplying intensive professional aid on a temporary basis. This is very significant to small-scale industries, which find it more economical than to employ full time senior staff, especially when the problems are occasional.
- Giving impartial viewpoints. In certain cases like job evaluation and standard setting, the worker's union sometimes feel the recommendations made by the company's industrial engineer is pro-management and they insist such studies which would affect the labor, be done by outsiders, In

Work Organization and Methods Engineering for Productivity.
DOI: https://doi.org/10.1016/B978-0-12-819956-5.00022-4

some cases the unions also employ their own similar consultants to work in tandem with the management appointed consultant.
- Providing the management with arguments that justify the predetermined measure.

22.4 Requirements of a consultant

- Technical know-how.
- Professional experience.
- Creativity and analytical ability.
- Diplomacy and tact.
- Effective communication and ability to put across ideas effectively.

22.5 Attributes of a consultant

1. He shall possess systematic knowledge and skill to perform functions with authority.
2. He shall have autonomy in his decisions and not controlled, supervised or directed by the client.
3. He is not subject to political ideologies or controls, but sees himself as being affected by public interest.
4. He is bound by ethics in the performance of his duties.
5. He will use his knowledge, skill and authority ethically.
6. He shall not knowingly do harm to his clients.
7. He shall not have bias in his judgments.

22.6 Qualities of consultants as per P.W. Shay

P.W. Shay in his book *The Common Body of Knowledge for Management Consultants* cites the qualities of consultants as follows:

1. Good physical and mental health.
2. Professional etiquette and courtesy.
3. Stability of behavior and action.
4. Self-confidence.
5. Personal effectiveness (drive).
6. Integrity.
7. Intellectual competence.
8. Good judgment.
9. Independence. He must be self-reliant but not subordinate to other's opinions. He must be able to form his own opinions in areas of his competence and the experience and at the same time recognize the limitations of his competence, experience and judgment.
10. Strong analytical or problem-solving ability.
11. Creative imagination.

12. Skill in interpersonal relationships.
13. Ability to communicate with above average facility.
14. Psychological maturity.

22.7 External and internal consultants

External consultants are administratively and legally independent of the organization for which they work except for the contractual obligations.

Internal consultants are generally employees on full time contract with monthly salary for a fixed period. They include internal auditors, legal consultants and industrial engineers. Their functions are similar to the consultants to the effect they are asked to study certain specific aspects and report. Here lies their need for detailed knowledge and effective communication. In fact, every engineer himself is an internal consultant in his company, whether conducting a method study program or maintenance planning as explained in Chapter 1, and hence needs to have independent thinking, and effective communication.

22.8 Consultants' responsibility to the clients

1. **Confidentiality:** The engineer as a consultant would be given access to certain confidential documents or processes or information as a part of his task. At the same time he is not in full term employment and he may work for some other client also. If the second company happens to be the competitors of the first, there are always pressures on him to leak the confidential information. As an ethical consultant it is his primary responsibility to ensure that he does not pass off any information nor make use of that information without the consent of the first company. This aspect with reference to the engineering employees of the company has been discussed in detail in the chapter on conflict of interests in this author's book on Professional Ethics.
2. **Codes of ethics:** The consultant shall understand and know all the relevant canons and codes developed by the respective association and follow them where possible. As an illustration, some of the codes developed by European Federation of Association of Management Consultants are cited in paragraph 22.9.
3. **Safety of the project:** Consulting engineers have the responsibility in decision-making concerning the safety. As an outsider, the consultant should use his own judgment and ethical thinking in incorporating safety in every aspect of the project, even though the client feels it is too expensive.
4. **Refusal to be part of the unethical process and practices of the client:** As explained above the consultant can always and must refuse to tow the

line of the client if he feels that the client is deliberately asking for inferior designs, just for reducing the cost of the project.

5. **Critical loyalty:** If their professional judgment is overruled where the safety, health, property or the welfare of the public is endangered, they shall notify the client or the employer or such authority as may be appropriate. They will take a similar action when they find that the codes of practices are violated by the client. Books on Professional Ethics may be referred to appreciate the appropriateness of this attribute to the management consultants.
6. **When the project is weak:** Consultants shall advise frankly when they believe the project would not be successful.
7. **Implementation of the project:** The consultant's duty does not end in developing and executing a project. He has the responsibility to ensure that the project is successfully implemented and maintained. This is similar to the method improvement studies done by industrial engineers, where the 7-step procedure (SREDDIM) is as given in Fig. 22.1. Here the last two viz, install and maintain are emphasized.
8. **Design only projects:** Even after completing the design only projects like building design project of an architect, he should continue to hold responsibility of the strength of the building even after the completion

FIGURE 22.1 SREDDIM, the 7-step procedure for methods improvement.

and occupancy, as long as the construction is as per his specifications. It is moral duty to check at random the quality of construction.

9. **Legal responsibilities:** As an independent on contract with the client, the consultant is subjected to many legal issues during the performance of his work. If anything goes wrong in the project, it is the consultant that would be questioned first. The consultant must realize this aspect before he accepts a contract.

22.9 Sample codes of ethics

The codes of ethics for the professional consultants framed by The European Federation of Association of Management Consultants (FEACO), are given in the annexure. Nevertheless, the do's and don'ts are reproduced here as an illustration.

All member associations must subscribe to FEACO's Code of conduct and practice under which it is regarded as unprofessional conduct

1. Do not advertise in a blatant and commercial manner.
2. Do not accept any trade commission, discounts or considerations of any kind in connection with the supply of services or goods to a client.
3. Do not have interest in firms supplying goods or services to their clients, or to be under their control, or to fail to make known any kind of interest likely to affect their service.
4. Do not calculate remuneration on any basis other than the agreed professional scale of fees.
5. Do not disclose confidential information regarding the client's activities.
6. Do not pay or accept payment for the introduction of clients except in accordance with the recognized and generally accepted professional practice in the country concerned.
7. Do not do anything that does not accord with the statutes of the profession.

22.10 Data vs information

As defined by David Bourgeois, an information system (IS) is an organized system for the collection, organization, storage and communication of information. The management information system (MIS) is such information compacted by an organization by the coordination, control, analysis, and visualization of information in an organization; for decision making. The consultant hence should understand all aspects of this MIS.

We use the terms data and information so often and so freely in our day to day life and invariably use them as synonyms. Nevertheless, they have slight conceptual differences. Data is the raw, unorganized facts that need to be processed. Data is something simple and apparently collected at random, and not much useful until it is organized. On the other hand, when data is processed, organized, structured or presented in a given context so as to make it useful, it is called information.

The website https://www.diffen.com/difference/Data_vs_Information distinguishes data and information as below.

22.11 Characteristics of management information systems

The following are the 9 factors that decide the effectiveness of MIS

1. Completeness
2. Clarity
3. Conciseness
4. Accuracy and reliability
5. Up-to-date
6. Relevancy
7. Simplicity
8. Predictability
9. Summarizing

Since the project reports originating from a work study engineer on the method study projects undertaken by them are akin to the MIS, these characteristics are further explained in Table 22.1.

22.12 Computerization of MIS

Today with so much information traveling around various persons, computerization becomes inevitable. In fact, Wikipedia refers MIS to the processing of information through computers and other intelligent devices to manage and support managerial decisions within an organization. They are also called transaction processing systems, decision support systems, expert systems, or executive information systems. The processor does the following functions

- Arithmetical calculations
- Sequential and repeated calculations
- Evaluation and comparison
- Choosing one or two or more alternatives
- Rearrangement of the data in the desired order
- Sending out the resulting signals to the output unit to the printer for data presentation.

Fig. 22.2 above illustrates the role of the computer program in data processing.

TABLE 22.1 Characteristics of data and information.

	Data	Information
Meaning	Data is raw, unorganized facts that need to be processed. Data can be something simple and seemingly random and useless until it is organized.	When data is processed, organized, structured or presented in a given context so as to make it useful, it is called information.
Etymology	"Data" comes from a singular Latin word, datum, which originally meant "something given." Its early usage dates back to the 1600s. Over time "data" has become the plural of datum.	"Information" is an older word that dates back to the 1300s and has Old French and Middle English origins. It has always referred to "the act of informing," usually in regard to education, instruction, or other knowledge communication.
Example	Each student's test score is one piece of data.	The average score of a class or of the entire school is information that can be derived from the given data.
Features	1. Data is a set of discrete, objective facts about events. 2. Symbols which represent information for processing purposes, are based on explicit or implicit agreements about the meaning of the data 3. Data is factual information (as measurements or statistics) used as a basis for reasoning, discussion, or calculation	1. Information is knowledge of ideas, facts and/or processes 2. Information is meant to change the way the receiver perceives something, to have an impact on his judgment and behavior 3. Think of information as data that makes a difference 4. Information is data interpreted in its original meaning.

22.13 Report writing and work study engineer

Report writing is a significant function a work study engineer, consultant or anyone working on a project. He may be highly qualified, may have extensive professional experience on the subject, may have creativity and analytical ability of the highest caliber, but he can succeed in his profession only if he can prepare effective reports and communicate to others effectively, whatever he conceives, develops and recommends. For this he must understand the principles of management information system, as explained in earlier paragraphs.

Way back in 1948, Harold Lasswell, a professor at Yale Law School, put forward a maxim also called Lasswell's communication model that describes communication by defining who said it, what was said, in what channel it was said, to whom it was said, and with what effect it was said.

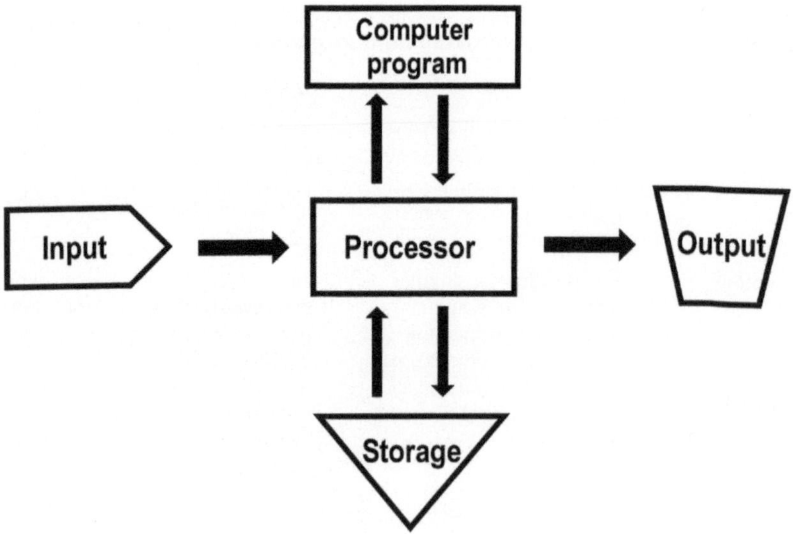

FIGURE 22.2 Data processing by computer.

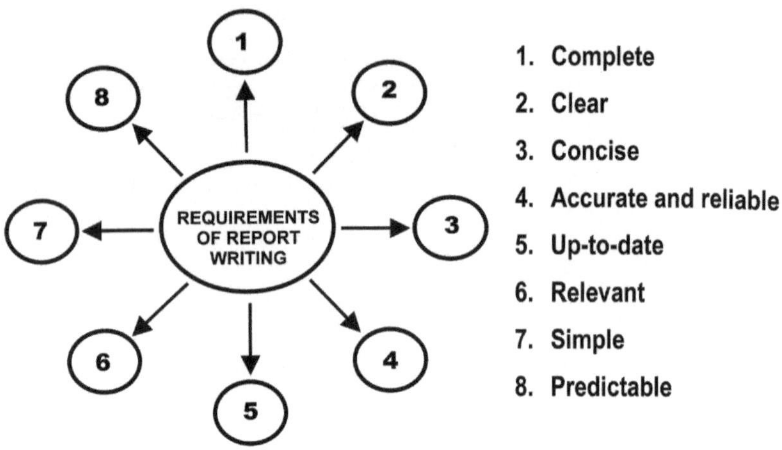

FIGURE 22.3 Requirement of a consultant's report.

In light of these points, this chapter also gives few tips on effective communication with reference to project reporting and presentation, as illustrated in Fig. 22.3.

1. **Complete:** The report must possess all the elements needed for the decision making and inform about all relevant factors for the change.
2. **Clear:** The report presented must be clear and without being vague to avoid wrong interpretations.

3. **Concise:** The report should be concise and preset complete picture without unnecessary descriptions with use of standard formats, figures, and symbols. This is akin to the Japanese terms Kanso, Shizen and Shibum, as explained further in chapter 22. 32
4. **Accurate and Reliable:** The report should be accurate and reliable and presented in such a manner as to avoid misunderstanding.
5. **Up-to-date:** The report must be updated frequently or periodically. All the old and obsolete information and figures must be deleted.
6. **Relevant:** The report should be relevant to the theme of the study and be trimmed off all the unwanted references and information.
7. **Simple:** The report should be prepared in simple and straight language, so as to be understood by all concerned without any misinterpretation.
8. **Predictable:** The report should assist in the management's capacity to predict or highlight the problems before they occur.

22.14 Basic steps of project report writing

The basic steps involved in any development project are

1. Collect data,
2. Assemble, classify and summarize data,
3. Record and present data,
4. Examine and analyze data,
5. Develop the recommendations your report.
6. Prepare the report and present it to the concerned authority, and
7. Implement the recommendations your report and help in maintaining the new procedures.

22.14.1 Data collection

Once a project is selected the first step is to collect data. This can be done by

- Direct observation methods,
- Interviewing,
- Referring top published records,
- Questionnaires.

22.14.2 Recording and presentation of data

In general, the collected data can be presented in variety of ways

- Narrative form
- Tabular presentation
- Graphical presentation
- Use of symbols and charts
- By figures using symbols, pictures and cartoons.

22.14.3 Forms of reports submitted

- Personal discussions
- Feasibility reports
- Final reports
- Periodic management reports

22.14.4 Tips for personal discussions

- Plan your talk based on
  - Your overall objective
  - The group
  - The time available
  - The resources available
- Plan and schedule the subject matter
- Understand the barriers of communication
- Encourage the audience participation

22.15 Basic communication skills

22.15.1 What is communication?

Communication is the exchange of facts, ideas, opinions or emotions by 2 or more persons, it is the broad field of human interaction and interchange of the above rather than the technology of the communication channels like telephone or radio.

22.15.2 Elements of communication

1. A communicator	(e.g.: the speaker or the report writer)
2. The act of communication	(e.g.: the report writing or explanation)
3. The channel	(e.g.: the verbal talk or written report)
4. Stimuli	(e.g.: emphasis or visual aids)
5. The communicatee	(e.g.: audience or the top management)
6. Response	(e.g.: reply or reaction)
7. Motivation	(e.g.: the purpose of the communication)

22.15.3 Barriers of communication

In spite of our best efforts, the message we send or the verbal explanation tend to be missed or wrongly understood by the communicatee. Summarizing of such difficulties or the 'barriers of communication' would help us in trying to avoid them as far as possible.

1. Badly prepared message
2. Talk unsuited to the audience

3. Dull delivery, lack of audio or visual stimuli
4. Inaudibility
5. Physical distance
6. Mannerisms
7. Faulty transmission
8. Unclarified assumptions
9. Hostility and distrust caused by bias and predetermined opinions and impressions
10. Lack of understanding
11. Personal preoccupation
12. Physical condition like comfort of the chairs, A/C, etc.
13. Neighbors
14. Passing distractions like cell phones ringing in the middle of a lecture
15. Emotional attitudes
16. Poor retention

22.16 Case study - consultancy requirements of a medium scale industry of Chennai in the Indian context

The following details would illustrate the several consultancy requirements of a Medium scale industry manufacturing screw jockeys as conducted by the author

1. NSIC & SIDCO applications
2. Organizational planning
3. Selection and recruitment of personnel
4. Counseling in labor management relations
5. Plant layout for the factory
   a. in the existing factory
   b. in the new building ($60' \times 40'$ shed) to be allotted by Small Industries Development Corporation (SIDCO)
6. Project planning & execution by PERT for the new factory up to pilot production
7. Process planning for each of the 10 products, each with about 4 sizes.
8. Design of operations
9. Machinery selection
10. Routing
11. Preparation of operation process charts
12. Preparation of process planning layouts
13. Setting up of quality standards
14. Design of gauges
15. Selection and specification of standard tooling
16. Design of jigs and special tooling
17. Establishment of production procedure and PP&C systems and formats

18. Setting up of stores and inventory control systems including fixation of stock levels.
19. If the new project is a modification of the existing process in a new location, Value analysis and methods improvement studies to be conducted on both the processes.
20. Setting up of standard times of all operations and production targets.
21. Systems manual for the whole organization.

22.17 Conclusion

Apart from the role discussed in Chapter 1, the work-study engineer assumes the role of an internal consultant in the organization in which he works by acting as an educator, a catalyst for deeper change, a resourceful facilitator, contributing with all the characteristics discussed in this chapter.

Criteria questions

1. How can you say a work-study engineer is an internal consultant of an organization? (22.1)
2. What are the attributes of a consultant? (22.5)
3. Discuss the responsibilities of a consultant to the organization which he works for. (22.8)
4. What do you understand by the code of ethics for the consultant? (22.9)
5. Distinguish between data and information. (22.11)
6. Discuss the requirements of report writing. (22.14)
7. Discuss the consultant's need for good communication skills. (22.15)

Further reading

1. ILO. *Introduction to work study*. 3rd ed. Geneva: ILO; 1979.
2. Barnes R. *Work sampling*. J. Wiley & Sons; 1961.
3. Longla CT. *Work sampling as an indicator of productivity, building trades*. 1987.
4. Salvendy G. *Handbook of industrial engineering: technology and operations management*. J. Wiley & Sons; 2001.
5. Kiran DR. *Professional ethics and human values*. 2nd ed. McGraw Hill Education; 2007.
6. Kiran DR. *Production planning and control: a comprehensive approach*. BS Publications; 2018.
7. Lambrou FH. *Guide to work sampling*. J.F. Rider. 1962.
8. Kiran DR. *Report on the strategy requirements for the development of a small scale manufacturing industry - a* report submitted to M/S Vanjax Industries, Chennai - April. 1993.

Work study syllabi from the Indian universities and professional bodies

1 Indian Institution of Industrial Engineering for the Grad. I.E Examinations

IEA 05: Work Systems Design

A. Concept & need of productivity at operator level in any activity-Inputs-Process Output relationship, Work Study approach in improving Productivity; Factors affecting productivity. Necessity of standard method; good working condition and measuring work content.

B. Method Study Definition, scope and application of method study as a logical approach to problem solving in a wide range of activities; Basic procedure Select, record, Analyse and Examine, Develop New Method, Install, Maintain Report to management - presentation, Discussion; Installation; Maintaining.

C. Work Measurement - Aims and objectives of work measurement, definitions of terms, tools, techniques.

- **(a)** Stop watch study: (i) Equipment used, procedure for conducting a study, (ii) Recording data-forms, timing methods, timing aids, (iii) Rating Concept- scales, methods; Allowances-Concept, types and calculations; Standard time, standard production; Comments on stop - watch study.
- **(b)** Activity Sampling - Definition, objectives scope of application; Statistical aspect-sampling, confidence level, accuracy, number of observations, use for determination of standard time, idle time, standard production.
- **(c)** Production study - definition, scope of application, procedure, limitations; Predetermined Time Standard systems - (i) Background of PTS (ii) Detailed study of MTM-1.

D. Job evaluation, Incentive system design, Theory of learning curve.

E. Ergonomics - Definitions and scope of ergonomics. Need of Fit the job Operator; Anthropometry - Definition and applications in work study; Environment Factors-noise, light heat vibration, humidity their sources, effect on human body, effect on productivity and corrective actions. Place of Ergonomics in work study and their interdependence.

2 Anna University

Subject - IE8303 Work System Design

Unit I: Productivity: Total time for a job or operation, total work content and ineffective time, - Production and Productivity - Productivity and standard of living, Factors affecting Productivity, Introduction to Productivity measurement Models.

Unit II: Methods Engineering: Methods Engineering-Steps -Tools and techniques, Motion study.

Unit III: Work Measurement: 9 Stop watch time study, performance rating, allowances, Development of Standard data, learning effect. Work measurement in Automated Processes. Computerised Labour standards.

Unit IV: Applied Work Measurement: Work sampling, Group Timing Technique (GTT), predetermined time systems, types, Methods Time Measurement (MTM), Introduction to MOST standard, Wage incentive plans.

Unit V: Work Design for Office Work: Organization and methods (O & M), Work measurement of office work, Work Analysis techniques applied to support staff, Form design and control.

3 Andhra University

Subject IE 101E: Methods Engineering and Work Design

Work Study: Concept of work and productivity - Possibility guides - Methods study - Charting techniques - Concept of standard time and bench mark jobs - Timing techniques and work sampling - Elemental motions, THERBLIGS and principles of motion - Economy - Introduction to predetermined motion time standards.

Human Factors Engineering: Introduction to ergonomics and human factors - Engineering physiological basis of human performance - Biomechanics - Psychology of work and work load perception - Physical work environment - Basis of ergonomic problem identification - Safety.

Integration of methods and time - Learning theory implications on standard time - Work study applications in production, maintenance, quality and other service functions - Synthetic time standards - MTM system and its application to production and maintenance.

Organization and Methods: Procedure analysis and developing office standards - MTM application to office work - Forms design and control - Records management.

Value Engineering: VE concepts, Principles, Methodologies and standards - Methods of functional analysis - Creativity - VE case studies/project work. Job evaluation and incentive scheme: Job description and job analysis.

Job evaluation- different methods - Individual and group incentive concepts and implications - Different types of incentive schemes - Suggestion schemes.

4 Andhra University

DDMEC-302 (Management courses): Industrial Engineering and Management

Unit I: Concepts of Industrial Management: Principles of management- Growth of management thought, Functions of management, Principles 3 of organization, Types of organisation and committees. Introduction to personnel management-Functions, Motivation, Theories of motivation, Hawthrone studies, Discipline in industry, Promotion, Transfer, lay off and discharge, Labour turnover, Industrial relations Trade unions, industrial disputes, Strikes, Lock-out, Picketing Gherao, Settlement of Industrial disputes, Collective bargaining, Industrial dispute act 1947 and factories act 1948.

Unit II: Production Planning and Control: Types of productions. Production cycle, Product design and development, Process planning Forecasting, Loading, Scheduling, Dispatching, Routing, Progress, Control, Simple Problems.

Unit III: Plant Layout: Economics of plant location, Rural Vs Suburban sites, Types of layouts, Types of building, Travel chart technique, Assembly line balancing simple problems. Materials Handling-Principles, Concept of Unit load, Containerisation, Pulletisition, Selection of material handling equipment, Applications of belt conveyors, Cranes, Forklift trucks in industry, Plant Maintenance-Objective and types.

Unit IV: Work Study: Concept of productivity (Simple problems), Method Study- Basic steps in method study, Process chart symbols, Charts, Diagrams and Models used, Principles of motion economy, Therbligs, Simo chart, Work Measurement-Stop watch procedure of time study, Performance rating and allowances, Work sampling, Simple problems.

Unit V: Materials Management: Introduction, Purchasing, Objectives of purchasing department, Buying techniques, Purchase procedure, Stores and material control, Receipt and issue of materials, Store records, Quantity Control - Inventory, Functions of inventory, Simple EOQ model, ABC analysis, Simple problems. Quality Control - Single and Double sampling plans. Control charts of variables and attributes (Use of formulae only).

5 Jawaharlal Nehru Technological University (JNTU) Hyderabad

B.Tech 4th Year 1 sem Mechanical Engineering

Industrial Management (Elective - I) R13 syllabus

Unit I: Introduction to Management: Entrepreneurship and organization - Nature and Importance of Management, Functions of Management, Taylor's Scientific Management Theory, Fayol's Principles of

Management, Maslow's Theory of Human Needs, Douglas McGregor's Theory X and Theory Y, Herzberg's Two-Factor Theory of Motivation, Systems Approach to Management, Leadership Styles, Social responsibilities of Management.

Unit II: Designing Organizational Structures: Departmentalization and Decentralization, Types of Organization structures - Line organization, Line and staff organization, functional organization, Committee organization, matrix organization, Virtual Organization, Cellular Organization, team structure, boundary less organization, inverted pyramid structure, lean and flat organization structure and their merits, demerits and suitability.

Unit III: Operations Management: Objectives- product design process-Process selection-Types of production system (Job, batch and Mass Production), Plant location-factors- Urban-Rural sites comparison- Types of Plant Layouts-Design of product layout- Line balancing (RPW method) Value analysis-Definition-types of values- Objectives- Phases of value analysis- Fast diagram.

Unit IV: Work Study: Introduction - definition - objectives - steps in work study - Method study - definition - objectives - steps of method study. Work Measurement - purpose - types of study - stop watch methods - steps - key rating - allowances - standard time calculations - work sampling. Statistical Quality Control: variables-attributes, Shewart control charts for variables- chart, R chart, - Attributes-Defective-Defect- Charts for attributes-p-chart -c chart (simple Problems), Acceptance Sampling- Single sampling-Double sampling plans-OC curves.

Unit V: Job Evaluation: Methods of job evaluation - simple routing objective systems - classification method - factor comparison method - point method - benefits of job evaluation and limitations. Project Management (PERTICPM): Network Analysis, Programme Evaluation and Review Technique (PERT), Critical Path Method (CPM), Identifying critical path, Probability of Completing the project within given time, Project Cost Analysis, Project Crashing (simple problems).

6 DR. B. R. Ambedkar National University of Technology, Jalandhar

B.Tech in Industrial and production Engineering

IPX 303 Work Study and Ergonomics

Work-Measurement: Definition, various techniques of work-measurement work-sampling, stopwatch time study & its procedure, Job selection, Equipment and forms used for time study, rating, methods of rating, allowances and their types, standard time, numerical problems, predetermined - time standards and standard data techniques.

Incentive: Meaning, objectives of an incentive plan, various types of incentive plans.

Course Contents

Section-A

Productivity: Definition, reasons for low productivity, methods to improve productivity, work-study and productivity

Human factor in work-study: Relationship of work-study man with management, supervisor & workers, qualities of a work-study man.

Method-study: Definition, objectives, step-by-step procedure, questioning techniques, charts and diagrams for recording data. Like outline process charts, flow process charts, multiple activity charts, two handed process chart, string diagram, travel chart, cycle graph, Chrono-cycle graph, therbligs, micro motion study and film analysis, Simo chart, principles of motion economy. Development and installation of new method.

Section-B

Work-Measurement: Definition, various techniques of work-measurement work-sampling, stopwatch time study & its procedure, Job selection, Equipment and forms used for time study, rating, methods of rating, allowances and their types, standard time, numerical problems, predetermined - time standards and standard data techniques.

Incentive: Meaning, objectives of an incentive plan, various types of incentive plans.

Section-C

Ergonomics: Introduction, history of development, man-machine system and its components. Introduction to structure of the body- features of the human body, stress and strain, metabolism, measure of physiological functions- workload and energy consumption, biomechanics, types of movements of body members, strength and endurance, speed of movements. NIOSH lifting equation, Lifting Index, Maximum acceptable Weights and Forces, Distal upper extremities risk factors, Strain Index, RULA, REBA. Applied anthropometry - types, use, principles in application, design of work surfaces and seat design. Visual displays for static information, visual displays of dynamic information, auditory, tactual and olfactory displays and controls. Assessment of occupational exposure to noise, heat stress and dust. Effect of vibration/ noise, temperature, illumination and dust on human health and performance.

7 National program on technology enhanced learning, - a joint initiative of Iits and Iisc

1. **Work Study**
   a. Historical background; Work study definition; Role of work study in improving productivity; Ergonomics and work study.
   b. Work study procedure: selection of jobs; Information collection and recording; Recording techniques -charts and diagrams; critical analysis; developing better method; installation and follow up of standard method.
2. **Motion Study**

   Memomotion and micromotion study; therbligs; cyclegraph and chronocyclegraph; simochart; Principles of motion economy; Design of work place layout.

3. **Work Measurement**

Definition; Procedure; Performance rating; Concept of normal time; allowances. Work sampling technique of work measurement. Introduction to pre -determined motion time system.

8 Amaravathi University

Subject: Work Study

1. What is Work Study? It is the systematic examination of the methods of carrying out activities such as to improve the effective use of resources and to set up standards of performance for the activities carried out. "It is one of the most important management techniques which is employed to improve the activities in production. The main objective of WS is to assist the management in the optimum use of the human & material resources." Main aim of this WS is to finding the best & most efficient way of using available resources i.e. men, material, money & machinery.
2. What is Work Study? "It is especially concerned with productivity. To increase the productivity from a given quantity of resources." Work study is a generic term for the techniques of method study and work measurement. These techniques are used in the examination of human work in all its contexts. They lead systematically to the investigation of all the factors which affect the efficiency and economy at the work place in order to affect improvement.
3. Components of Work Study. It has three aspects 1. More effective use of plant & equipment 2. More effective use of human effort 3. Evaluation of human work.
4. Applications of Work Study: Industries: production operations, research and development. Marketing: sales & distribution. Offices: stores & warehouses. Material handling in designing field. Building & other construction, Transport, Hospital, Army, Agriculture.
5. Advantages/Objectives of work Study.
6. Method Study Method Study Flow Chart "Method study is the technique of systematic recording and critical examination of existing and proposed ways of doing work and developing an easier and economical method." Method study examines the way a task (changing the clutch on a car, preparing a flower bed for planting, cleaning a hotel room) is done. The industrial engineer has an eye on operational efficiencies and costs, quality of processes, service reliability, staff safety, etc. Method study techniques are applicable from factory/workshop manufacturing to cabin crew activities.
7. Basic Procedure for Method Study Method Study Flow Chart The basic procedure for conducting method study is as follows: 1. Select the work

to be studied. 2. Record all facts about the method by direct observation. 3. Examine the above facts critically. 4. Develop the most efficient and economic method. 5. Define the new method. 6. Install the new method. 7. Maintain the new method by regular checking.

8. Basic Procedure for Method Study: 1. Select: While selecting a job for doing method study, the following factors are considered: (a) Economical factors. (b) Human factors. (c) Technical factors. 2. Record: All the details about the existing method are recorded. This is done by directly observing the work. Symbols are used to represent the activities like operation, inspection, transport, storage and delay. Different charts and diagrams are used in recording. They are: 1. Operation process chart 2. Flow process chart 3. Two-handed process chart 4. Multiple activity chart 5. Flow diagram 6. String diagram
9. Basic Procedure for Method Study: 3. Examine: Critical examination is done by questioning technique. This step comes after the method is recorded by suitable charts and diagrams. The individual activity is examined by putting a number of questions. 4. Develop: Develop the most efficient and economic method. 5. Define: Once a complete study of a job has been made and a new method is developed, it is necessary to obtain the approval of the management before installing it. The work study man should prepare a report giving details of the existing and proposed methods.
10. Basic Procedure for Method Study: 6. Install: This step is the most difficult stage in method study. Here the active support of both management and trade union is required. Here the work study man requires skill in getting along with other people and winning their trust. 7. Maintain: The work study man must see that the new method introduced is followed. The workers after some time may slip back to the old methods. This should not be allowed. The new method may have defects. There may be difficulties also. This should be rectified in time by the work study man.
11. Objectives of Method Study: 1. Improvement of manufacturing processes and procedures. 2. Improvement of working conditions. 3. Improvement of plant layout and work place layout. 4. Reducing the human effort and fatigue. 5. Reducing material handling. 6. Improvement of plant and equipment design. 7. Improvement in the utility of material, machines and manpower. 8. Standardization of method. 9. Improvement in safety standard.

9 Rajiv Gandhi Technical University (RGTU)

ME 701 (D) Work Study and Ergonomics Syllabus

Unit 1: Method Study: Purpose of work study, its objectives, procedure and applications; method study definition and basic procedure, selection of job, various recording techniques like outline process charts, flow process charts, man machine charts, two handed process charts, string diagram, flow diagram, multiple activity chart, simo, cyclographs and chrono-cyclographs; critical examination, development, installation and maintenance of improved method; principles of motion economy and their application in work design; micro motion study, memo motion study and their use in methods study.

Unit 2: Work Measurement: Introduction & definition, objectives and basic procedure of work measurement; application of work measurement in industries; time study: basic procedure, equipment needed, methods of measuring time, selection of jobs, breaking a job into elements; numbers of cycles to be timed; rating and methods of rating, allowances, calculation of standard time. Work sampling: Basic procedure, design of work sampling study conducting work sampling study and establishment of standard-time.

Unit 3: Job Evaluation and Incentive Schemes: Starlight line, Tailor, Merrick and Gantt incentive plans Standard data system; elemental and non-elemental predetermined motion systems, work factors system; Methods Time Measurement (MTM), MOST.

Unit 4: Human Factor Engineering: Definition and history of development of human factors engineering, types & characteristics of man-machine-system, relative capabilities of human being and machines; development and use of human factor data; information input and processing: Introduction to information theory; factors effecting information reception and processing; coding and selecting of sensory inputs.

Unit 5: Display Systems and Anthropometric Data: Display- types of visual display, visual indicators and warning signals; factorial and graphic display; general principles of auditory and tactral display, characteristics and selection.

10 B.M.S. College of Engineering (Autonomous), Bengaluru-19

15IM4DCIEG INDUSTRIAL ENGINEERING:

Unit 1: Productivity: Definition of productivity, individual enterprises, task of management Productivity of materials, land, building, machine and power. Measurement of productivity, factors affecting the productivity, productivity improvement programs, wages and incentives (simple numerical problems).

Unit 2: Work Study: Definition, objective and scope of work study. Human factors in work study. Work study and management, work study and supervision, work study and worker. **Method Study:** Definition, objective and scope of method study, activity recording and exam aids. Charts to record moments in shop operation - process charts, flow process charts, travel chart and multiple activity charts. (With simple problems).

Micro and Memo Motion Study: Charts to record moment at work place - principles of motion economy, classification of moments two handed process chart, SIMO chart, and micro motion study. Development, definition and installation of the improved method, brief concept about synthetic motion studies.

Unit 3: Work Measurement: Definition, objective and benefit of work measurement. Work measurement techniques. Work sampling: need, confidence levels, sample size determinations, random observation, conducting study with the simple problems. **Time Study:** Time Study, Definition, time study equipment, selection of job, steps in time study. Breaking jobs into elements, recording information Rating & standard Rating, standard performance, scale of rating, factors of affecting rate of working, allowances and standard time determination; Predetermined motion time study - Method time measurement (MTM)

Unit 4: Ergonomics: Introduction, areas of study under ergonomics, system approach to ergonomics model, man-machine system. Components of man-machine system and their functions - work capabilities of industrial worker, study of development of stress in human body and their consequences. Computer based ergonomics.

Design of Man-machine System: Fatigue in industrial workers, Quantitative qualitative representation and alphanumeric displays, Controls and their design criteria, control types, relation between controls and displays, layouts of panels and machines. Design of work places, influence of climate on human efficiency. Influence of noise, vibration and light.

Unit 5: Current Trends: Introduction to Agile manufacturing, Lean and Six Sigma, Value Engineering, Just in time, Total quality management, Enterprise resource planning, Supply chain and logistics management.

Industrial Engnineering Laboratory

I. **Method Study** (1) Recording Techniques using charts. (i) Outline process chart. (ii) Flow process chart. (iii) Multiple Activity Chart. (2) Recording Techniques using diagrams. (i) Flow diagram. (ii) String diagram. (3) Application of principle of motion economy: (i) Two handed process chart. (4) Assembling simple components. (i) Peg board assembly. (5) Development of Plant Layout using: (i) Black board

II. **Work Measurement** (1) Rating practice using: (i) Walking simulator. (ii) Pin board assembly. (iii) Dealing a deck of cards. (iv) Marble collection activity. (2) Determining the standard time using: (i) Simple operations using stopwatch time study. (ii) Predetermined Method Time Study (PMTS).

III. **Ergonomics** (1) Measurement of heart beat rate, calorie consumption parameters using: (i) Walking simulator. (ii) Cycle Ergo-meter. (2) Effect of human efficiency in work environments: (i) Noise. (ii) Light.

11 Thiagarajar College of Engineering

M.E. DEGREE (Industrial Engineering) PROGRAMME

14IE130 Work Study and Cost Analysis

Syllabus:

Productivity and Work Study: Productivity and standard of living, Techniques to reduce work content and ineffective time. Productivity matrix, Quality route to productivity, better asset utilization, wages and salary, job evaluation, job description, job analysis and merit rating, Leveraging IT for improved productivity - Case studies. Work Study - Introduction - Human factors.

Method Study: Introduction - Selection of jobs - Recording - Tools and Techniques - Charts, Diagrams, Template and Models - Examining - Developing the improved method - Principles of motion economy. Work Measurement: Introduction to Work Measurement - Time study equipments - Selecting the job to be studied and making a Time Study- Rating - Allowances to Standard Time - Setting Time Standard for work with machines - Examples of time study. Other Techniques of work measurement - Production study - Activity Sampling - Synthesis - Analytical Estimating - Predetermined Motion Time Systems. The use of Time standards - Organization of a work study department.

Ergonomics: Psycho physiological Data - Anthropometry, information displays - Man Machine System - Working Environment -chair and table heights. Strength and force of body movements - speed and accuracy of motor responses.

Activity Based Costing: Definition - Purpose - cost estimation Vs cost accounting - components cost - Direct cost - indirect cost - overhead expenses. Estimation of cost elements - set up time and economic lot size - tool change time - Inspection time - performance factors - overheads. different methods of apportioning overheads - Data required for cost estimating -Steps in making a cost estimate - estimation of production cost of simple components - problems.

Summarized syllabi – foreign universities

King Abdulaziz University, Jeddah, Saudi, IE 341

1. **Introduction to Work Study and System:** The Nature of Work, Work Study, Important of Work Study and Principles, Defining Work Systems, Types of Occupations, Productivity, Excess Nonproductive Activities, Allocation of Total Task Time.
2. **Methods Engineering & Layout Planning:** Methods Engineering Definition & objectives, Operations Analysis, Methods Analysis, Methods Design, Techniques of Methods Engineering: Charting & Diagramming Techniques, Motion Study and Work Design. Objectives of Charts and Diagrams, Charts and Diagrams: Operation Charts, Process Charts (Flow Process Chart, Worker Process Chart & Form Process Chart), Flow Diagram, Activity Charts (Right hand/left-hand activity chart, Worker-Machine Activity Chart, Worker-multimachine activity chart). Basic Motion Elements, 17 Therbligs Basic Motion Elements, Micro-motion Analysis, Principles of Motion Economy.
3. **Time Study and Work Measurement:** Time Standards definitions and How They Are Determined, Time study definitions. Functions of Time Standards, Methods to Determine Time Standards, Allowances in Time Standards. Direct Time Study Procedures, Performance Rating, Time Study Equipment. Overview of Predetermined Motion Time Systems, Methods-Time Measurement (MTM), MTM-1, MOST. Using a Standard Data System, Work Element Classification in Standard Data Systems. Work Sampling Defined, Work Sampling Applications, Statistical Basis of Work Sampling, Number of Observations Required, Use of Work Sampling to Measure Average Task and Standard Times. Learning Curve Theory, Determining the Learning Rate, Crawford model in learning curve, Time Standards Versus the Learning Curve.

American University of Beirut

INDE 320: Work Measurement and Methods Engineering
Course topics:

1. Importance of Productivity and Work Design
2. Problem-Solving Tools
3. Operations Analysis and Lean Manufacturing
4. Work Design
5. Standards and Time Studies
6. Work Sampling
7. Predetermined Time Systems
8. Uses of Standards: Costs and Wages

North Eastern University, Boston

Course number: MIM U310 Introduction to Industrial Engineering
Course topics:

- History of Engineering and Industrial Engineering: Development and Scope
- Manufacturing Engineering and Operations Planning
- Facilities Location and Layout
- Material Handling, Distribution and Routing
- Work Design and Work Measurement
- Quality Control
- CAD/CAM. Robotics and Automation
- Human Factors
- Financial Management and Engineering Economy
- Operations Research
- Simulation and Queuing Systems
- Project Management
- Lean Manufacturing, Six Sigma, Supply Chain Management

City University of Hong Kong

Department of Systems Engineering and Engineering Management
Course no. SEEM3034 Work Design
Keyword syllabus

1. Productivity Measurement and Indices.
2. Standards and Work Methods Design.
3. SREDDIM Approach of Method Study. Principles of Motion Economy.
4. Time Study Techniques: Performance Rating and Allowances.
5. Standard Data and Formulas.
6. Work Sampling.

7. Learning Curve.
8. Multi-factor Wage Incentive Plans.
9. Job Shop and Flow Shop.
10. Assembly Line Balancing.

Index

Note: Page numbers followed by "*f*" and "*t*" refer to figures and tables, respectively.

F

T

Y

FSC
www.fsc.org
MIX
Papier | Fördert
gute Waldnutzung
FSC® C083411

Kolibri 360 GmbH
Ferdinand-Jühlke-Straße 7
99095 Erfurt, Deutschland
produktsicherheit@kolibri360.de